Introductory Biophysics

Introductory Biophysics

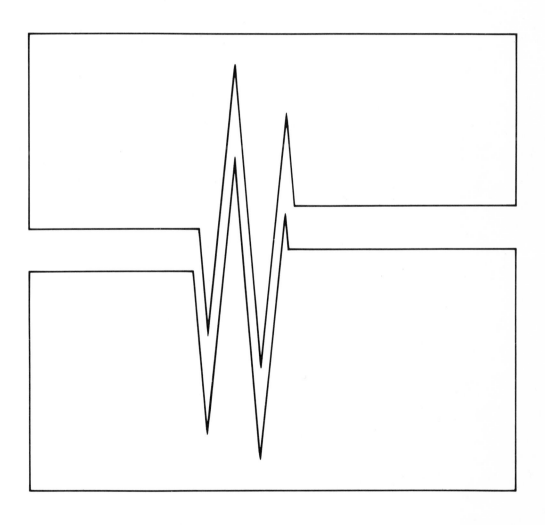

Introductory Biophysics

F. R. Hallett
University of Guelph

P. A. Speight
University of Toronto

R. H. Stinson
University of Guelph

in association with

W. G. Graham
University of Guelph

A Halsted Press Book
John Wiley & Sons
New York

Published in the U.S.A. by Halsted Press, a division of John
Wiley & Sons, Inc., New York.

Library of Congress Cataloging in Publication Data
Hallett, Frederick Ross, 1942–
  Introductory biophysics.
  Bibliography: p.
  Includes index.
  1. Biological physics. I. Speight, P.A., joint
author. II. Stinson, Robert Henry, 1931– joint
author. III. Title
QH505.H25   574.1'91   76-21809
ISBN 0-470-15195-1

QH
505
H25
1977

Printed and bound in the United States of America
1 2 3 4 5   81 80 79 78 77

Contents

Preface

As is quickly noted from glancing through the chapters in this book, there is a considerable amount of physics and mathematics. We make no apologies for this because we are convinced that one cannot hope to really understand the operation of a biological process such as, for example, vision without understanding in some detail pertinent physical phenomena. Further, it will be apparent that the book places a very heavy emphasis on problem solving. This was done with the conviction that the process of working problems really does help the student to come to grips with important concepts. The level of the problems is such that they can be successfully tackled by first- or second-year university students with a limited background in physics and mathematics. While the problems do not involve calculus, the calculus is used in some formula derivations. A great deal of emphasis is placed on simple wave functions (sine and cosine) and on functions of the form $N/N_0 = e^{kx}$ because of their biological importance in population growth and decay, in light absorption, in radiation-biophysics, in bioenergetics and in spike propagation along an axon. It is likely that many students will have to proceed through the exercises in the appendices before beginning Chapter 2.

Our goal, then, is to demonstrate that one can gain good insight into a biological process by approaching it from the physical point of view. Our experience is that a course of this type can be enjoyed by biology, agriculture and physical science students alike, especially if they have the opportunity to interact.

Acknowledgements

The authors wish to thank Dr. G. H. Renninger, Dr. J. L. Hunt, Dr. B. M. Millman, Dr. R. S. Gage, Mr. E. McFarland of the Department of Physics of the University of Guelph and Dr. D. A. Pink of the Department of Physics of St. Francis Xavier University for their useful comments and critical evaluation of many sections of this text.

Chapter 1
What Is Biophysics?

The natural scientists of previous centuries are often remembered for the broad diversity of their areas of interest. Helmholtz, for example, was noted for his extensive contributions to thermodynamics, optics, electromagnetism and for his theories and experiments on hearing and vision. Their work not only provided the foundations for the modern scientific disciplines of biology, chemistry, and physics, but encouraged the development of numerous subdisciplines as well. One of these subdisciplines, biophysics, grew very slowly until the 1940s. In North America, the development of nuclear devices and the accompanying high energy radiation spawned the rapid growth of radiation biophysics. The interest in radiation biophysics continued to grow, encouraged by the development of radiation therapy of cancer and the puzzling effects of radiation on cells. In the mid-fifties, when the Biophysical Society was established, a large fraction of its members were associated with radiation biophysics. In Europe, Britain especially, the growth of biophysics took a different turn, primarily because of the remarkable successes of x-ray diffraction studies of biological macromolecules. Many biophysicists were, and still are, attracted to this area of research. The main stumbling block to the development of other areas in biophysics was the lack of quantitative and analytical tools suitable for the study of biological systems. In the fifties, many

of the needed tools became available. Of particular importance is the computer. It has given the biophysicist the ability to model extremely complicated biological systems and monitor the effects of many parameters.

In our opinion, biophysics is the study of the physical aspects of biological structure and function. The tools of the trade are the techniques and theoretical methods of the physical sciences. Biophysics removes the traditional barriers encompassing the disciplines of physics, chemistry, biology, and medicine. It flourishes where interdisciplinary communication abounds.

Until recently, those embarking on a future in biophysics had no formal training in the subject. Invariably they had studied one or more of mathematics, engineering, biology, or medicine and had considerable experience before moving to biophysical problems. Delbruck, Kendrew, von Békésy, Crick, Meselson, Hartline, Gamow, Schrödinger, Hodgkin, Huxley, and many others are examples of this trend. The success of their approach is unquestioned; many in fact, became Nobel Prize winners. Along with their successes the body of knowledge in biophysics has mushroomed. Today graduate and undergraduate courses in the subject can be found all over the world. The subject is fascinating to us; we hope that it is to you.

1

Chapter 2
Sound, Hearing and Echolocation

2-1 Introduction

In 1793, an Italian scientist, Spallanzani, wrote of many of the experiments he had performed on the navigation of bats around obstacles. His interest in the problem had been stimulated when he noticed that his pet owl crashed into walls when his candle was extinguished, whereas the bats in his room maintained flight and avoided obstacles successfully. The conclusions from his work and that of many of his contemporaries, Jurine especially, was that the hearing process of the bat was somehow responsible. Still, their skepticism was great. Referring to experiments where bats had successfully avoided fine wires stretched across a darkened room he said, "But how, if God love me, can we explain or even conceive in this hypothesis of hearing."

Today we know considerably more about the process, called echolocation. It is but one of the many fascinating topics related to sound and its perception.

2-2 The Nature of Sound

Sound is a wave disturbance which propagates whenever atoms or molecules are present. In fact, it is the motions of these particles which constitute sound. As is true for other types of waves, there is no net transfer of matter as the wave progresses; the particles oscillate back and forth about their equilibrium positions.

These oscillations are extremely small. When we investigate hearing we will find that sounds which are faint, but easily heard, cause displacements in air molecules of fractions of a nanometre. The sound wave is a longitudinal wave which means that the oscillations are parallel to the direction of propagation of the wave.

It is important to study some of the characteristics and properties of sound waves, because these in turn help in understanding the processes of hearing and echolocation. The student may at certain times need to review the appendix on waves and vibrations before proceeding.

2-3 Energy, Power, and Intensity of a Sound Wave

Three closely related properties of a sound wave are energy, power, and intensity. Energy is transported in the wave although, as we've mentioned earlier, no transfer of matter occurs. An exact macroscopic analogy of a sound wave is impossible to produce, but the following crude picture may help.

Imagine, first of all, that air molecules act like small balls connected by weightless springs. If, by some trickery they were lined up (as in Figure 2-1) and then the end ball was struck, each ball in turn would move, and pass on the motion to the next ball. A disturbance would travel down the string of balls and the

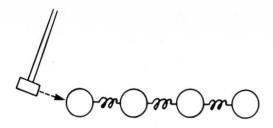

Figure 2.1

Mechanical Analogy of a sound wave

one at the other end would be projected. If the string of balls was arbitrarily long, the disturbance would be transmitted throughout the entire length without any of the balls experiencing a net change in position. A sound wave is a good deal more complicated than this crude model. Air molecules fill a volume so that the sound disturbance is transferred by whole groups of molecules. Also, the wavelength of sound is much greater than the size of an air molecule. The displacement of groups of molecules is given by the equation

$$y = y_o \sin \left(\frac{2\pi x}{\lambda} \right) \qquad (2\text{-}1)$$

where y_o is the maximum displacement or displacement amplitude of the molecules, x corresponds to distance along the wave, and λ is the wavelength of the disturbance. This function is shown in Figure 2-2. Remember as you look at this figure that the actual displacement of the groups of air molecules is parallel to x, the direction the wave is travelling.

Figure 2-2 is really a snapshot at a particular instant of time of the function which corresponds to the travelling sound wave. The travelling wave, including time dependence (t) is given by

$$y = y_o \sin \left(\frac{2\pi t}{T} - \frac{2\pi x}{\lambda} \right) \qquad (2\text{-}2)$$

This equation is discussed in detail in Appendix 3. It can also be written (Appendix 3, problem 2):

$$y = y_o \sin 2\pi\nu\left(t - \frac{x}{v} \right) \qquad (2\text{-}3)$$

where ν is the frequency of the wave and v is the propagation velocity of the wave. Table 2-1 gives a selected list of sound propagation velocities in several media.

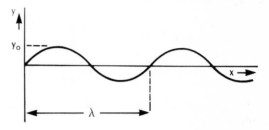

Figure 2.2

Displacement of air particles as a function of distance from the source of sound

Table 2-1
The Velocity of Sound in Various Media

Medium	Approximate Sound Velocity (m s^{-1})
air	340
water	1550
iron	5000
rubber	54
wood (pine)	3300

The energy of the sound wave is transported along by the oscillating motions of the molecules of the carrier substance. If we sum the energies of each of the particles then we can obtain an expression for the total amount of energy in the wave. The instantaneous kinetic energy of one of the molecules is

$$E_{Ki} = \tfrac{1}{2}mv_i^2 \qquad (2\text{-}4)$$

where m is the mass of the molecule and v_i is its instantaneous velocity. Note that v_i is the instantaneous velocity of the particle and not the propagation velocity of the wave, which is v. We can obtain an expression for v_i by recognizing that

$$v_i = \frac{dy}{dt} \qquad (2\text{-}5)$$

That is, the instantaneous velocity is the time rate of change of the particle displacement. Using Equation (2-3)

$$v_i = \frac{d}{dt} y_o \sin 2\pi\nu\left(t - \frac{x}{v}\right)$$

$$= y_o 2\pi\nu \cos 2\pi\nu\left(t - \frac{x}{v}\right) \qquad (2\text{-}6)$$

From Equation (2-4) the instantaneous kinetic energy E_{Ki} is

$$E_{Ki} = 2\pi^2 m y_o^2 \nu^2 \left[\cos 2\pi\nu\left(t - \frac{x}{v}\right)\right]^2$$

$$(2\text{-}7)$$

The maximum value of the kinetic energy occurs when

$$\cos 2\pi\nu\left(t - \frac{x}{v}\right) = 1 \qquad (2\text{-}8)$$

At this instant the potential energy of the particle is zero so that the maximum kinetic energy is equal to the total energy E_p of the particle. This is because energy is continually converted back and forth from kinetic to potential in all forms of simple harmonic motion, such as wave motion, vibration and pendulum motion. The most obvious example is a pendulum such as the one shown in Figure 2-3. As it swings by its equilibrium position, its velocity is a maximum, its kinetic energy is a maximum, but its potential energy is zero. On the other hand, the instant the pendulum is maximally displaced from its equilibrium position, its velocity is zero, its kinetic energy is zero, but its potential energy is a maximum. At all times the total energy of the system is

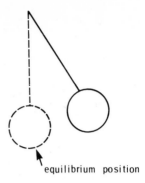

equilibrium position

Figure 2.3
Simple pendulum

the sum of the potential and kinetic energy terms. Thus at the instant Equation (2-8) holds

$$E_p = \text{maximum kinetic energy}$$

$$= 2\pi^2 m y_o^2 \nu^2 \qquad (2\text{-}9)$$

This energy, E_p, is the total energy of one of the particles in the medium through which the sound wave is propagating. For a gas or liquid of density ρ, the energy per unit volume, E_ρ/V present in the wave is given by

$$\frac{E_\rho}{V} = 2\pi^2 \rho y_o^2 \nu^2 \qquad (2\text{-}10)$$

The total energy present in a particular wave is then

$$E_\rho = 2\pi^2 \rho \nu^2 y_o^2 V \qquad (2\text{-}11)$$

where V is the total volume of the wave. Notice that the amount of energy in a wave is proportional to the square of the sound frequency, the square of the particle displacement amplitude and the density of the medium.

EXAMPLE

Typical echolocating pulses (Figure 2-4) from a resting bat have a frequency of about 5×10^4 Hz and a duration of 60 milliseconds.

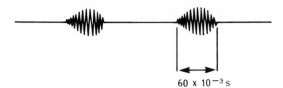

60×10^{-3} s

Figure 2.4
Orientation pulses from a resting bat (Adapted from "Discrimination of Thin Wires by Flying Horseshoe Bats (Rhinolophidae)" by H. U. Schnitzer in "Animal Sonar Systems" N.A.T.O. Advanced Study Inst.)

The density of air is about 1.3 kg m^{-3} and the velocity of sound in air is 340 m s^{-1}.

(a) How long is the sound wave associated with each pulse?
(b) If the energy per cubic meter of the pulse near the bat is 2.5×10^{-9} J m^{-3} what would be the corresponding displacement amplitude of air molecules in the wave?

Length of wave = velocity of sound × duration of the pulse

$$= 340 \times 60 \times 10^{-3}$$
$$= 20.4 \text{ m}$$

The energy per unit volume of the wave is related to the wave parameters by Equation (2-10).
Thus

$$y_o^2 = \frac{(E_\rho/V)}{2\pi^2 \rho v^2}$$
$$= \frac{2.5 \times 10^{-9}}{2 \times 9.9 \times 1.3 \times (5 \times 10^4)^2}$$
$$= 3.9 \times 10^{-20} \text{ m}^2$$
$$y_o = 2.0 \times 10^{-10} \text{ m}$$
$$= 0.20 \text{ nm}$$

Imagine that a sound wave is coming your way. You know that energy is being transmitted from the sound source to you. However, you may also want to know how much energy per unit time is carried by the wave. This quantity, energy transmitted per unit time, is called power, P. Thus

$$P = E_\rho/t \qquad (2\text{-}12)$$

where t is time.

The power carried by the wave is, of course, equal to the power produced by the sound emitter assuming no losses. The units of power are joules per second or watts.

EXAMPLE
What acoustical power is produced by the bat when it emits a 2.5 millisecond pulse having a total energy of 2.5×10^{-9} joules?

$$P = E_\rho/t$$
$$= \frac{2.5 \times 10^{-9} \text{ J}}{2.5 \times 10^{-3} \text{ s}}$$
$$= 10^{-6} \text{ watts}$$

One of the most useful parameters describing a wave is the intensity. Intensity, I, corresponds to the amount of power per unit area carried by the wave. The area, A, referred to here is the area through which or onto which the sound wave is passing. Thus

$$I = \frac{P}{A} = \frac{E_\rho}{At} \qquad (2\text{-}13)$$

The acoustical power produced at a typical source is shown schematically in Figure 2-5.

Assuming that the air around the source absorbs or attenuates the sound negligibly, then the total amount of power passing through an imaginary sphere O must be identical to the total amount of power passing through imaginary sphere Q. The sound intensity at Q is less than at O because the same power is spread over a larger area. Since the area of any sphere is $4\pi r^2$, the intensity at any distance r is

$$I = \frac{P}{4\pi r^2} \qquad (2\text{-}14)$$

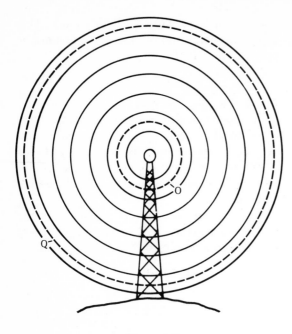

Figure 2.5

Sound wave radiating from point source

This is called the inverse square relationship and is found in many areas of physics.

EXAMPLE

A foghorn is used to warn ships that dangerous rocky shoals are nearby. The intensity of the foghorn at 1 metre is 10^{-3} W m^{-2}. At sea, a sound intensity of about 10^{-8} W m^{-2} is required for a man to hear the device. How far away from the horn can it safely be heard?

In general

$$I = \frac{P}{4\pi r^2}$$

so that

$$P = 4\pi r^2 I$$

and

$$r^2 = \frac{P}{4\pi I}$$

At 1 m

$$P = 4\pi \times 10^{-3} \times (1)^2$$
$$P = 4\pi \times 10^{-3} \text{ watts}$$

At maximum distance r

$$10^{-8} = \frac{4\pi \times 10^{-3}}{4\pi r^2}$$
$$r^2 = 10^5 \text{ m}^2$$
$$= (3.2 \times 10^2 \text{ m})^2$$
$$r = 3.2 \times 10^2 \text{ m}$$

Sound intensity can also be related to the displacement amplitude (y_o) of the oscillating particles of the medium. The intensity, which is the amount of energy striking a unit area per second, is the energy per unit volume E_ρ/V of the wave times the velocity (v) of the wave. Thus

$$I = \left(\frac{E_\rho}{V}\right)v = 2\pi^2\rho\nu^2 y_o^2 v \qquad (2\text{-}15)$$

There is another, more common way of expressing intensities called the intensity level, L. The intensity level is actually a means for describing intensities relative to a standard intensity. Although the intensity level is dimensionless, it is, nevertheless, quoted in bels, or decibels (db).

Thus

$$L = \log (I/I_o) \qquad (2\text{-}16)$$

where L is the intensity level in bels, I_o is the reference intensity and I is the intensity in question. If L is expressed in decibels (db) then

$$L = 10 \log (I/I_o) \qquad (2\text{-}17)$$

For sound in air I_o is conventionally taken as 10^{-12} W m^{-2} and

$$L = 10 \log (I/10^{-12}) \text{ db.} \qquad (2\text{-}18)$$

Table 2-2 shows several common sounds and their intensity levels. It is interesting to note that this scale arose because of the human sensory response to sound levels. If the background sound (I) is low then only a small change (ΔI) in sound intensity is required for us to respond to, or hear the change. However, if the background is large, then a larger change in sound intensity is required to produce the same response. It has been found experimentally that

$$h = \frac{\Delta I}{I} \qquad (2\text{-}19)$$

where h is a measure of the human response. The equation applies to most sensory responses and is called the Weber-Fechner law. Equation (2-19) has the form of the derivative of a logarithmic function such as Equation (2-16).

Table 2-2
Intensity Level of Common Sounds

	Intensity Level (db)
Ear drum ruptures	160
Pain threshold	120
Shotgun blast	100
Car horn at 6 metres	90
City street	75
Conversation	55
Typical room	40
Open country	10
Threshold	0

EXAMPLE

The intensity of a bat's pulse a few centimetres from the bat is about $10^{-2}\,\mathrm{W\,m^{-2}}$. Express this in db.

$$L = 10 \log \frac{10^{-2}}{10^{-12}}$$

$$= 10 \times 10 = 100 \text{ db.}$$

Note from Table 2-2 that this intensity level is similar to that obtained from a shotgun blast. Fortunately for us the frequency of the cry is outside the range of the human ear.

2-4 The Ear and Hearing

The ear is one of the most remarkable devices known to science. Its range of sensitivity is unparalleled, extending roughly 12 orders of magnitude from the threshold of hearing to the threshold of pain. At 3.4 kHz, the threshold of hearing is about $10^{-12}\,\mathrm{W\,m^{-2}}$, an intensity which is only slightly greater than the "sounds" due to random thermal collisions of air molecules with the eardrum. Figure 2-6 shows the threshold of audibility over the frequency range 10 Hz to 9 kHz. Actually, small children can hear frequencies as high as 40 kHz, but the upper frequency drops to about 20 kHz by the time they reach the mid-teens. From there the upper frequency limit drops about 80 Hz every six months so that by the time a person is fifty, the limit is down to 12–15 kHz.

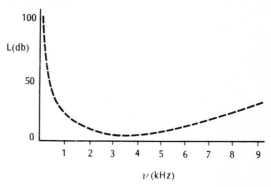

Figure 2.6
Threshold of audibility for the human ear

The ability to determine pitch or frequency of sound is another highly developed property of the human ear. The author remembers a particularly embarrassing event of his university years when in rehearsal of a particularly loud section of Handel's Messiah, the conductor suddenly stopped the whole 100 voice choir, pointed his baton and said "You sang that note $\frac{1}{4}$ tone flat." The note in question was a "D" of frequency 587 hertz (Hz). That conductor was capable of hearing the author singing at a frequency of about 10 Hz less than this or 577 Hz in the midst of 99 other voices.

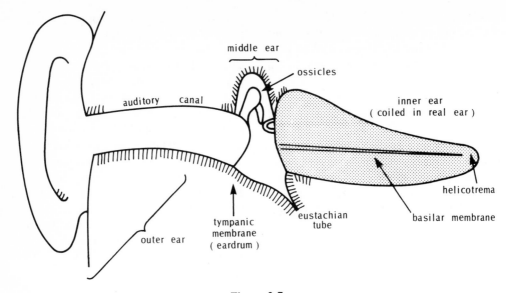

Figure 2.7

Schematic diagram of the human ear

In the following sections we will consider several of the properties and mechanisms of the ear. For more complete discussions of the subject the student is wise to search for the works of Georg von Békésy, one of the foremost experts on the subject.

Some of the mechanical and structural components of the ear are shown in Figure 2-7. Because of function as well as tradition, the ear is divided into three parts designated simply as outer, middle, and inner ear. Their functions are roughly as follows:

Outer ear —optimization and amplification
of the important
frequencies
Middle ear—transduction and amplification
Inner ear —pitch resolution and detection

Many of the sounds which occur around us are more important than others. The ability to hear the speech sounds of other humans is sociologically important as well as being important for our survival. The outer ear plays a large role in helping us hear the frequencies associated with this kind of sound while at the same time reducing the effect of less important sounds. The mechanism by which this is achieved, is in part, the phenomenon of resonance. In sound, this resonance arises from the production of standing waves in the auditory canal. The effect is somewhat similar to the production of standing sound waves in an open-ended organ pipe. In the appendix on waves and oscillations, we considered standing waves on a string. In the case of the auditory

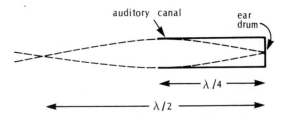

Figure 2.8

Fundamental wavelength of tube closed at one end

canal or open-ended organ pipes, the situation is similar. At the closed end of the tube, which is the eardrum, there must be a node. The fundamental resonant frequency occurs when the length of the canal is equal to one quarter of the corresponding wavelength. This is shown in Figure 2-8. Since the auditory canal is about 2.5 cm long, the optimum wavelength for resonance is about $4 \times 2.5 = 10$ cm $= 0.1$ m. The corresponding sound frequency can be obtained from the equation

$$\nu = v/\lambda = \frac{340}{0.1} = 3400 \text{ hertz.}$$

It may be observed from Figure 2-6 that we are maximally sensitive to sounds of about this frequency. It is interesting to note that a baby's cry is in this frequency range, a clever trick of nature to make sure that our attention is quickly obtained. Other animals may require optimization at other frequencies and the dimensions of their outer ears are changed accordingly.

The middle ear has two functions. First of all it must respond efficiently to the incoming sound disturbance. This is no simple task because sound waves are easily reflected at surfaces. In order to "couple" effectively with the sound the eardrum has to move with the air molecules as they oscillate at the sound frequency. To do this, the ear drum has to be extremely elastic and delicate. It is, therefore, easily ruptured.

Some idea of the displacement of the eardrum by sound can be obtained by applying Equation (2-15) at the threshold of hearing and the threshold of pain. We will use a frequency of 4 kHz which is in the optimum discernible frequency range illustrated by Figure 2-6.

At the threshold of hearing at 4 kHz, the sound intensity is about 10^{-12} W m^{-2}. The displacement amplitude, y_o, of the air molecules gives us a lower limit to the displacement of the eardrum. Using Equation (2-15)

$$y_0 = \sqrt{\frac{I}{2\pi^2 \rho \nu^2 v}}$$

$$= \sqrt{\frac{10^{-12}}{2\pi^2 \, 1.3(4 \times 10^3)^2 \, 340}}$$

$$= 2.7 \times 10^{-12} \text{ m}$$

$$= 2.7 \times 10^{-3} \text{ nm}$$

This distance, which is about $\frac{3}{100}$ of the diameter of a hydrogen atom, is astoundingly small.

Pain occurs when sound intensities reach about 120 db, a value which corresponds to about 1.0 W m^{-2}. At this intensity,

$$y_o = \sqrt{\frac{1.0}{2\pi^2 \, 1.3(4 \times 10^3)^2 \, 340}}$$

$$= 2.7 \times 10^{-6} \text{ m}$$

$$= 2.7 \times 10^3 \text{ nm}$$

This is still a remarkably small distance. It is approximately the shortest length which can be seen with the best light microscope.

The combined effect of the eardrum and the ossicles, the small bones of the middle ear, is to convert the sound energy into mechanical energy. The lever action of these ossicles leads to more forceful displacement at the oval window although a reduction in amplitude simultaneously takes place. The process is analogous to the operation of a pressure intensifier as shown in Figure 2-9, where the oil at pressure P_1 on the area A_1 of the piston exerts a force

$$F_1 = P_1 A_1$$

This force is transmitted through the piston of area A_2 to oil at pressure P_2. Since

$$A_1 P_1 = A_2 P_2$$

then

$$P_2 = (A_1/A_2)P_1$$

and the pressure is amplified by the factor A_1/A_2.

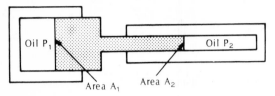

Figure 2.9
A pressure intensifier

The details of the operation of the inner ear are by no means as well understood. Figure 2-10 shows a highly schematicized diagram of this section of the ear with the cochlea uncoiled. Essentially the organ contains two fluid filled chambers divided by a membrane, called the basilar membrane. This membrane does not, however, completely separate the two chambers. A small opening, called the helicotrema, is present at the distal or innermost part of the system. The basilar membrane contains the neural detecting cells, which are small hair-like cells which are triggered by shear effects occurring between the fluid and the membrane. Thus, any lateral or sideways motion of the membrane with respect to the fluid is detected. The oval window is forced to vibrate by the action of the ossicles which, in turn, causes an oscillating pressure in the upper chamber. This pressure can

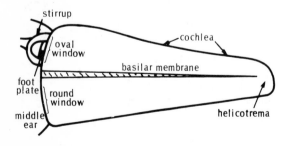

Figure 2.10
Diagram of the inner ear

be equilibrated or relieved by the membrane bulging downward as in Figure 2-11. The bulging process produces the shearing effects which the hair cells detect. The nervous system takes it from there.

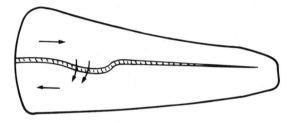

Figure 2.11
Distortion of the basilar membrane

The basilar membrane is fairly thick and stiff near the oval window, but tapers continuously to a thin elastic membrane near the helicotrema. This type of construction is thought to be important in the determination of the pitch of the sound. High frequency disturbances are quickly damped out and cause bulging only in the first, thick part of the membrane. For very low frequencies of around 50 Hz, the whole membrane bulges. For sounds lower than 30 Hz the disturbance goes off the end of the membrane, causes liquid to flow through the helicotrema and hearing becomes difficult. While this description does offer some basis for pitch discrimination, many scientists feel that this mechanism alone cannot account for the astonishing pitch resolution of the human ear. They believe that lateral interactions or cross-talk between auditory neural receptors is also involved. This is sometimes called neural sharpening.

In concluding our discussion of the hearing process, we should point out that the fine details of its operation are often more complicated than we have described. For example, resonance effects in the auditory canal are important in determining our optimum hearing range, but other factors are involved because our sensitivity falls off more slowly at higher or lower frequencies than the "pure" theory would predict. We've neglected as well to discuss pressure in our description of sound transduction by the eardrum. In fact, at low frequencies the pressure oscillations are very important in the hearing process. This is evidenced by the observation that when the eustachian tube connecting the middle ear to the environment is plugged as in a cold, we lose our ability to hear low frequency sounds and things sound strange.

Finally, we've made no attempt to discuss the various psychological factors involved in our conscious ability to listen to certain sounds or voices and reject others. An understanding of these functions will have to await great improvements in our knowledge of the operation of the brain.

2-5 Echolocation by Bats

It's near midnight; no moon; the leaves and the grass are rustling gently in a light breeze. Every now and then a bat sails by, a flitting shadow. The general quiet is so profound and refreshing that it is difficult to believe that a regular cacophony of high frequency sounds are present. These sounds, which fortunately we cannot hear, are sound pulses issued several times per second by bats as they locate obstacles and their lunch, flying insects. In terms of intensity these pulses are roughly equivalent to those which occur in shotgun blasts. In terms of efficiency, the bat can easily catch several hundred flying insects per hour, relying solely on his sophisticated echolocating skills.

Modern research on echolocation is extensive, especially because of potential applications in the military. Consequently a large amount of this type of research is classified and unavailable to us. Nevertheless, many excellent books and papers on the subject can be

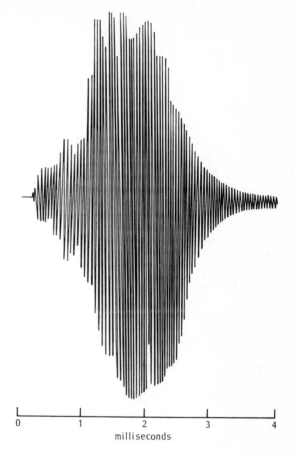

Figure 2-12
Echolocating pulse of a long-eared bat (from "Listening in the Dark" by D. R. Griffin, New York: Dover Publications, Inc. 1974)

found. The work of Donald R. Griffin especially forms the basis of much of the field.

In the following discussion we will mention two types of echolocating processes used by bats. The first uses short frequency modulated pulses. The second uses long continuous pure tones.

The shape and duration of a typical short pulse used by the long-eared bat is shown in Figure 2-12. The pulse is both amplitude modulated (AM) and frequency modulated

(FM) so that it has some similarities to radio communication. The modulation serves many purposes. Under some conditions, such as flying in caves, the bat has to recognize which echoes are his in the midst of other echoes from his fellows. In addition, the modulation must somehow give the bat information on the size, shape, motion, and perhaps even texture, of the object. The FM property of the pulse is thought to be particularly important here. To understand this in more detail one should first note that the frequency of a typical pulse descends from about 70 kHz to 30 kHz over its length. (The frequency change is what is meant by the term frequency modulation.) The student can quickly verify that this corresponds to a wavelength range of about 5–11 mm. The wavelengths which are equal to or less than the length of the object are scattered most effectively. Thus, if the object is small, a mosquito for instance, the returning pulse or echo will have a relatively larger component at the short wavelength region compared to the outgoing pulse. If the insect is flapping its wings then the effective dimensions of the bug change from one echo to the next. This changing character of the echoes presumably provides a clue to the bat that he is, indeed, tracking something tasty. This and other interesting aspects of insect capture have been considered in the various research papers of K. D. Roeder. Still unexplained, however, are the details of the process by which certain fish-eating bats can locate their prey from several feet above the surface. Also hard to explain is the ability of the bat to distinguish between circular and cross-shaped obstacles or circular and square-shaped obstacles. The amount of information present in the echoes is indeed remarkable.

Obviously a bat has to wait for the echo of one pulse before he can send another. Otherwise he would have difficulty associating the echo with the appropriate pulse. When freely flying and searching for insects he sends out pulses at a rate of about 4 per second. Each pulse has a duration of about 1.5×10^{-3} s. The firing rate can give us some idea of the maximum range of his echolocating system. The total distance travelled by the pulse and the echo is twice the distance (d) between the bat and the object. One can obtain d from a simple rearrangement of the expression for the pulse velocity. That is

$$d = \frac{vt}{2} = \frac{340 \text{ m s}^{-1} \times 0.25 \text{ s}}{2}$$

$$= 43 \text{ m}$$

However, it is doubtful that the bat's system is really capable of this large a range unless the object were as large as a building. For obstacles such as trees and posts the range is more like fifteen metres and for flying insects, only a couple of metres. We will analyse this point in more detail when we consider the intensity of the echo compared to the hearing threshold of the bat.

If, during this free flight, the bat detects an echo from what might be an insect the firing rate increases and the duration of each pulse decreases. If the object is indeed an insect and the bat chases it, then the firing rate increases to as high as 200 pulses s^{-1} with a duration of less than 10^{-3} s per pulse. At this firing rate the bat-object distance must be less than

$$d = \frac{vt}{2} = \frac{340 \times \frac{1}{200}}{2}$$

$$= 0.85 \text{ m}$$

The bat rarely catches the bug directly in his mouth. Usually he grabs the bug with a wing or foot and scoops it into his mouth. Many insects sense the acoustic pulses from the bat and begin evasive flight patterns. Sometimes these loops and power dives work, but usually the bat wins. He has the advantage of speed.

A slightly more sophisticated treatment of

the bat's echolocation range can be obtained by considering the intensity of the echo in relation to the threshold of hearing. The concepts involved are illustrated in the following example. The equations and definitions used have been discussed earlier in this chapter.

EXAMPLE

What is the maximum distance that a bat can detect an object of cross-sectional area of 10^{-4} m^2 if (i) the intensity of the pulse 10^{-2} m from the bat's mouth is 10^{-2} W m^{-2}, and (ii) the threshold of hearing of the bat is about 10^{-12} W m^{-2}, a value close to that for humans?

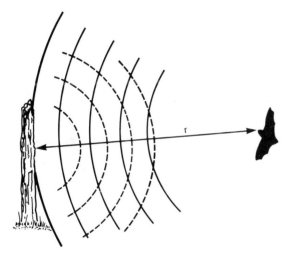

Figure 2.13

Detection of an obstacle by a bat

Using Equation (2-14) the power output of the bat can be evaluated. Thus

$$P = 4\pi r^2 I$$
$$= 4\pi (10^{-2})^2 10^{-2}$$
$$= 4\pi \times 10^{-6} \text{ watts.}$$

If this output power is assumed to be radiated in all directions as in Figure 2-13, then the power striking the object is given by

$$P_1 = \frac{a}{A} \times P$$

where a is the area of the object, and A represents the area of the sphere outlined in Figure 2-13 whose radius, r, is equal to the distance between the bat and the object. Thus

$$P_1 = \frac{10^{-4} \times 4\pi \times 10^{-6}}{4\pi r^2}$$
$$= \frac{10^{-10}}{r^2}$$

Referring again to Figure 2-13, one can see that the object now becomes the "source" of the echo. If we assume that all of the power striking the source is reflected then the intensity of the echo at a distance, r, from the object (which is the distance back to the bat) is given by

$$I_1 = \frac{P_1}{4\pi r^2}$$

In order for the bat to hear the echo I_1 must be at least 10^{-12} W m^{-2}. Therefore,

$$10^{-12} = \frac{10^{-10}}{4\pi r^4}$$
$$r^4 = \frac{10^2}{4\pi}$$
$$r = 1.7 \text{ metre}$$

Using this technique one can obtain values of r for objects of various size. Many bats can increase the distance r by emitting the echolocating pulse in a directed beam instead of all directions as in this example.

The echolocating mechanism which we have been discussing so far is common to many types of bats. There is another kind of echolocating system called CF or constant frequency system, such as is used by C. Parnelli. The pulse from these bats is often relatively long, about 0.1 s, with a small amount of frequency modulation at the beginning and the rest a long continuous pure tone at about 60.0 kHz. With pulses this long it is apparent that an echo will return to the bat before the outgoing pulse is completed. Initially, this puzzled many researchers because it was difficult to understand how the bat could hear a faint echo while he was still blasting out the pulse. Subsequent experiments such as those by Pollak, Henson, and Novick have demonstrated, however, that this bat is not particularly sensitive to sounds at 60.0 kHz, but very sensitive to sounds at about 61.8 kHz. The sensitivity curve or audiogram for this bat is shown in Figure 2-14 and the large increase in sensitivity at 61.8 kHz is apparent. On the basis of this audiogram it was clear that in this type of bat another physical phenomenon, the Doppler effect, was being used. The Doppler effect refers to the effective change in sound frequency which occurs when either the source or the observer or both are in motion. It is the effect which causes a change in the

pitch of the sounds from a train as it passes you. When approaching, the pitch is higher than would be found if the train was at rest. When moving away the pitch is lower than the rest value.

Figure 2.15

Source and observer at rest

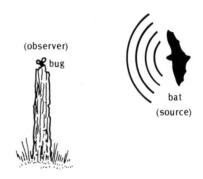

Figure 2.16

Source moving toward the observer

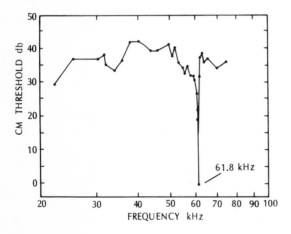

Figure 2.14

Audiogram from C. Parnelli. The threshold is expressed in db relative to the threshold at the best frequency (61.8 kHz). (from G. Pollak, O. W. Henson Jr. and A. Novick: Science 176, 66 (1972))

Let us first consider the Doppler effect using a pure tone source, an observer and the simple equation $\lambda = v/\nu$ relating sound velocity (v), frequency (ν), and wavelength (λ). If both the source and the observer are motionless with respect to each other, as in Figure 2-15 then we have

$$\lambda = \frac{v}{\nu} \qquad (2\text{-}20)$$

ν is the number of wave fronts per second arriving at the observer from the source. If the source moves toward the observer then it still produces sound at the same frequency, but the wavelengths are crammed into a smaller space as in Figure 2-16.

If the velocity of the source is v_a then λ_a (wavelength approaching) is given by

$$\lambda_a = \frac{v - v_a}{\nu} \qquad (2\text{-}21)$$

The frequency of the tone heard by the observer is

$$\nu_a = \frac{v}{\lambda_a}$$

or

$$\nu_a = \frac{v}{v - v_a} \nu \qquad (2\text{-}22)$$

Note that for this case, when the source is approaching the observer, the frequency he hears (ν_a) is slightly greater than the frequency produced by the source (ν). Similarly one can show that if the source is receding from the observer then the frequency the observer hears (ν_r for source receding) is

$$\nu_r = \frac{v}{v + v_r} \nu \qquad (2\text{-}23)$$

Note that in this case ν_r is slightly less than ν.

Now let's consider what happens when the source is at rest and the observer is moving. In this case the wavelength of the sound in the air is unchanged. If the observer is moving toward the source he encounters the wavefronts at a faster rate than if he were still. The frequency he hears then is slightly higher than he would have noticed had he been at rest. That is

$$\nu_a' = \frac{v + v_a'}{\lambda} = \frac{v + v_a'}{v} \nu \qquad (2\text{-}24)$$

where ν_a' is the frequency the observer detects as he approaches the source with a velocity v_a'. This state of affairs is shown in Figure 2-17. If

Figure 2.17

Observer moving toward the source

the observer were receding from the source then we would have obtained

$$\nu_r' = \frac{v - v_r'}{v} \nu \qquad (2\text{-}25)$$

Let's see how these equations work in analysing echolocation by bats using the CF system. While it's not exactly right, let's assume that the insect's motion is essentially zero with respect to the bat's motion. This is not too bad an assumption since bats fly considerably faster than insects. Thus at first the bat (source) is moving toward the insect (observer). The frequency the insect receives and therefore reflects is

$$\nu_a = \frac{v}{v - v_a} \nu$$

If the bat flies at a velocity equal to $5\,\mathrm{m\,s^{-1}}$ then

$$\nu_a = \frac{340}{340 - 5} (60 \times 10^3)$$

$$= 60.9\ \mathrm{kHz}$$

Regarding the echo, the insect is the source at 60.9 kHz and the observer, now the bat, is moving towards it. In this case

$$\nu_a' = \frac{v + v_a'}{v} \nu$$

$$\nu_a' = \left(\frac{340 + 5}{340}\right) \times 60.9 \times 10^3$$

$$= 61.8\ \mathrm{kHz}$$

The frequency heard by the bat, then, is exactly in his most sensitive region. He will home in on the insect because if he flies the wrong way the frequency shifts and the sound apparently fades because he is not quite as sensitive to it. As long as he maximizes the sound, the bat will encounter another snack. The wingbeat of the insect may modulate the echo frequency and provide clues to the nature of the object. If these modulations are absent the bat may recognize that the echo is from a solid object which must be avoided. A lot of this is speculation, however. A final interesting characteristic of many bats is that they limit the direction of the outgoing pulse to a beam of about 20 degree divergence (Figure 2-18). As it flies, this bat swings his head from side to side projecting his beam and scanning for echoes. The process reminds one of the rotating radar assembly visible at most airports.

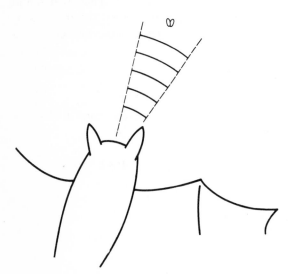

Figure 2.18
Scanning beam of the horseshoe bat

There are, as we have seen, many fascinating aspects to the echolocation problem.

There are many other related topics which are also interesting, but would require too much space to deal with adequately. One of these topics is echolocation by porpoises. Most libraries have excellent books on the subject, however, and the interested student is well advised to read them.

Problems

1. Conditions are such that a sound wave travels in air at $350 \, \mathrm{m \, s^{-1}}$. If its frequency is $10 \, \mathrm{Hz}$, what is the wavelength, the period, and the angular frequency of this wave?

2. In its search for flying insects, the bat uses an echolocating system based on the use of pulses of high frequency sound. Typically these pulses are about 2 ms long and have an intensity level of about 100 db at one metre from the bat's mouth. (Assume that the acoustical power is emitted in all directions.)

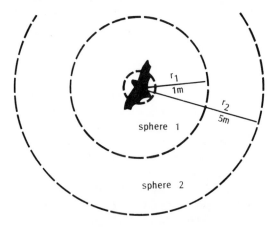

(a) What is the acoustical power associated with each pulse?
(b) How much acoustical energy is produced during each pulse?
(c) If the acoustical power is emitted radially in all directions, how much

power would go through the imaginary spheres of radius r_1 and r_2 shown in the figure?

(d) What would be the intensity of the pulse at r_1 and at r_2?

(e) Imagine a June Bug at a distance 5 m from the bat. The effective area of the June Bug is 10 mm². How much of the bat's acoustical power is intercepted by the bug? How much acoustical energy does he intercept from each pulse?

(f) What is the displacement amplitude of the air molecules at r_2 assuming that the frequency of the pulse is 5×10^4 Hz.

3. You are about to snatch a bright red strawberry in a supermarket when the store manager yells at you from 10 m away. The word he uses has a length in time of 0.2 seconds and a frequency of 500 hertz. His voice power during the word is 2×10^{-4} watts.

(a) How much acoustical energy actually struck your ears (total area 2×10^{-2} m²) and made you jump?

(b) What was the maximum displacement of your eardrum?

4. In moderate conversation, the human voice might radiate 10^{-5} watts. If there are no other sounds in the universe and if there are no losses of acoustical power in the air, at what distance would you have to be from a person speaking in such a way that the intensity of the sound of that voice would be at the threshold (assumed to be 10^{-12} W m⁻² at the frequencies present in the voice)?

5. The length of the auditory canal in a young human is approximately 1.3×10^{-2} m. What is the wave length and frequency of the fundamental standing wave in such a canal?

6. Whales are supposedly sensitive to low frequency under water sounds. If a hypothetical whale is sensitive to sounds in the 100 hertz region, suggest a reasonable length for his auditory canal.

7. A bat produces a sonar pulse with intensity of about 2×10^{-2} W m⁻². Express this in db.

8. The intensity of a particular sound is doubled. By how much will the intensity level change?

9. The sensitivity of the human ear is such that it can just detect a sound having an intensity of 10^{-12} W m⁻² at about 3000 Hz. Will the ear be able to hear a sound of (a) 30 db, (b) −10 db?

10. You are camping in mid-June when suddenly large cumulus clouds pile up and a thunderstorm strikes. With a "stroke" of good fortune you happen to have a sound level meter handy. On the second thunder clap you record a reading of 100 db which corresponds to a sound intensity of 10^{-2} W m⁻². This thunder clap reached you 4 seconds after the lightning flash. What was the acoustical power produced in the thunder clap (assume a point source)?

11. In the jargon of musicians, very loud and very soft sounds are described as "triple-forte" (fff) and "triple piano" (ppp). If fff corresponds to 10^{-2} W m⁻² and ppp is 10^{-8} W m⁻², express these in decibels.

12. Dogs have very sensitive hearing. Suppose their threshold of hearing is 10^{-15} W m⁻². If a sound is judged by a human to be 50 db what is the correct db rating for a dog?

13. A porpoise sends an echolocating pulse (60 kHz) as it tracks the path of a shark. The power associated with the pulse is 3×10^{-2} watts. The intensity of the pulse at the position of the shark is 1.5×10^{-5} W m⁻².

(a) What is the distance between the shark and the porpoise?

(b) What is the displacement amplitude

of the water molecules adjacent to the *shark*?

14. A CF bat emitting cries at 80 kHz flies directly at a wall. The frequency it hears is 83 kHz. How fast is it flying?

15. An FM bat emits a bunch of chirps. It detects an echo 5 ms after the chirps begin. How far away is the reflecting object?

Chapter 3
Optics of Vision

3-1 Introduction

The human eye is an extremely sensitive receiver of light. In order to understand how the eye functions it is necessary to investigate the nature of light, the behaviour of optical systems and the anatomy of the eye. The cellular and molecular aspects of the visual process will be discussed in Chapter 5.

3-2 The Nature of Light

Light is an electromagnetic wave phenomenon. Light waves constitute only a small fraction of the electromagnetic spectrum of waves shown in Figure 3-1. An electromagnetic wave differs from an acoustical wave in that no particle or material motion is involved. In fact, an electromagnetic wave can propagate perfectly well in a vacuum. The wave arises from closely coupled oscillations of magnetic (B) and electric (E) fields as illustrated in Figure 3-2. Note that the wave is a transverse wave with E and B oscillating in directions which are mutually perpendicular. The wave travels with a speed of about 3×10^8 m s^{-1} in a

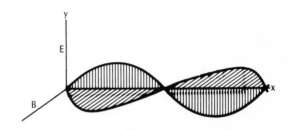

Figure 3.2
An electromagnetic wave

vacuum. This value, which seems to be one of the few genuine constants of our universe is usually given the symbol c.

We should at this point mention how these waves are produced. The question was clarified in 1885 in a series of experiments by Hertz. One might say, that the device which he used was, in a crude way, the world's first radio. The apparatus was composed of incomplete loops of wire and a high-voltage generator. One loop of wire was connected to the high-voltage generator and became the transmitter; the other the receiver. These components are illustrated in Figure 3-3.

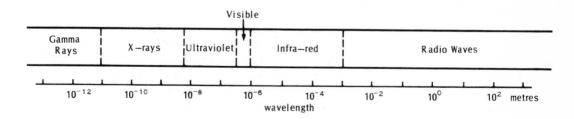

Figure 3.1
The electromagnetic spectrum

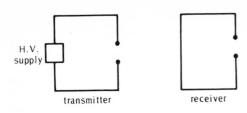

Figure 3.3
Hertz apparatus

When the high voltage supply was turned on charge piled up on the brass spheres of the transmitter as shown in Figure 3-4. When the potential difference between the spheres was sufficiently large, the insulating character of the air gap would be overcome and a spark, or surge of current would take place. If the voltage supply was constant, a sequence of firings or sparks would be observed. If the receiver was properly positioned with respect to the transmitter, Hertz noticed that a spark crossed its air gap at what seemed to be the same instant the spark crossed the transmitter's air gap. Yet the transmitter and receiver were not

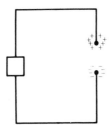

Figure 3.4
Building charge on the transmitter

connected in any way (Figure 3-5). He concluded that energy was carried from the transmitter to the receiver by an electromagnetic wave. Further experiments demonstrated that the only way one could get a wave was to cause a current surge such as a spark. In such surges, charge is accelerated and it is this acceleration of charge which causes the elec-

tric field and magnetic field disturbances characteristic to an electromagnetic wave.

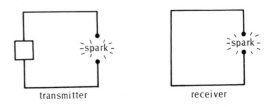

Figure 3.5
Sparks appear simultaneously on transmitter and receiver

As a final remark on the production of electromagnetic waves, it should be pointed out that there is no need for a wire. The wave could have been generated from electrons accelerating in space, in atoms, or in molecules. Thus, atoms and molecules can be considered to be transmitters or receivers of electromagnetic radiation as well.

3-3 Energy, Power and Intensity of Light

Light exhibits a wave-particle duality. That is, the propagation of light is best explained in terms of a wave phenomenon; but the interaction of light with matter is best explained in terms of light as a particle (photon). Since light is a wave it has a characteristic frequency (ν) and wavelength (λ) which are related by the expression

$$c = \lambda \nu \qquad (3\text{-}1)$$

This same wavelength and frequency are associated with the photon concept.

The energy of a particular electromagnetic wave is quantized into photons, each photon having an energy E which is proportional to the frequency of the wave. Thus

$$E = h\nu \qquad (3\text{-}2)$$

is the energy of one such photon. The constant in this equation, known as Planck's constant, has a value of 6.63×10^{-34} J s. These photons are indivisible in the sense that absorption of a photon by a molecule is an all or nothing process. A molecule can never absorb a fraction of a photon although a photon can be absorbed and subsequently a new photon can be emitted, at a different frequency.

Making use of the relationship between the speed of light, c, and the wavelength and frequency of the wave, the energy of a photon can be written as

$$E = \frac{hc}{\lambda}. \qquad (3\text{-}3)$$

Thus, while the photon energy is directly proportional to frequency, it is inversely proportional to wavelength. Since energy, wavelength, and frequency are related in this special way, one of these quantities automatically determines the other two. Thus, a particular electromagnetic wave may be characterised by energy, wavelength, or frequency. The dismaying result of this fact is the ridiculous number of units regularly used.

The most common means of describing the wave is probably in terms of wavelength. The internationally preferred unit of wavelength for UV and visible light is nanometres (10^{-9} m) but a few crusty old scientists refuse to be swayed from units like millimicrons (10^{-9} m) and Angstroms (10^{-10} m). Any other unit of length can be used, however, and often is. Units of frequency, cycles per second (hertz), are sometimes used but mostly in the microwave and radio-frequency part of the spectrum. The unit favoured by most molecular spectroscopists is wavenumbers which are found by first finding the wavelength in centimetres and then taking the inverse of that quantity (that is $1/\lambda$). The units of wave numbers are, therefore, cm^{-1}. As is obvious from Equation (3-3) these units are directly proportional to the photon energy.

EXAMPLE

What is the wavelength and frequency of a photon whose energy is 10^{-19} joules?
(a) The energy of a photon is 10^{-19} joules. The wavelength of this is

$$\lambda = \frac{hc}{E} = \frac{6.63 \times 10^{-34} \times 3 \times 10^8}{10^{-19}}$$
$$= 2 \times 10^{-6} \text{ metres}$$
$$= 2 \times 10^4 \text{ Angstroms}$$
$$= 2000 \text{ nanometres}$$
$$= 2000 \text{ millimicrons}$$

(b) The frequency of this photon is

$$E/h = 1.5 \times 10^{14} \text{ hertz}$$

In wavenumbers this corresponds to

$$1/\lambda \text{ (cm)} = 5000 \text{ cm}^{-1}$$

The power of a light source is expressed as the energy produced per unit time. Thus if a source gives off n photons of the same frequency in time t the power of the source is

$$P = \frac{nh\nu}{t} \qquad (3\text{-}4)$$

where $h\nu$ is the energy of each individual photon.

The intensity of light is the amount of energy per unit time passing through unit area and thus is power divided by the area A or

$$I = \frac{P}{A} \qquad (3\text{-}5)$$

As with sound, the intensity of light decreases as one over the square of the distance from the source to the receiver, that is

$$I \propto \frac{1}{r^2} \qquad (3\text{-}6)$$

3-4 Optical Elements of the Eye

Often the eye is likened to a camera, for its purpose is to form an image on the retina in the same way that a camera forms an image on a photographic plate. We have seen that light is a wave phenomenon but for a large device like the eye where the dimensions of the parts involved in bending the light are large compared to the wavelength of visible light (500 nm), the wave nature of light can be usually ignored and the light can be treated as travelling in straight lines or rays. This ray treatment of light is called geometrical optics and is a very good approximation.

Figure 3-6 shows a simple cross-sectional view of the human eye. Most of the refraction (bending) of light occurs at the cornea of the eye. The opening in the iris, the pupil, expands and contracts in response to the amount of light available. The primary purpose here,

as the camera enthusiasts will realize, is to maintain the light-gathering power at an optimum level for visual contrast and acuity (distinctness of the image). Some further refraction of the light takes place at the lens of the eye. However the main purpose of the lens is to allow the eye to focus over a wide range of distances, a process called accommodation. This is accomplished by a shape change of the lens effected by the action of the ciliary muscles. The retina on the back surface of the eye contains the photoreceptors (light sensitive elements) which convert the optical image into neural signals which are sent to the brain via the optic nerve. The fovea, a depression in the retina, is the site of day and colour vision. The photoreceptors here have a conical shape and are called cones. The rest of the retina contains mainly rod-shaped receptors which account for night vision and are colour blind. The region of the retina where the optic nerve

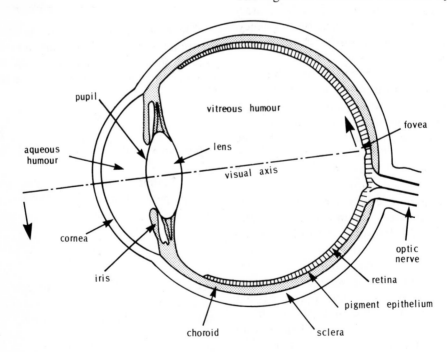

Figure 3.6
Lateral cross-section of the human eye

enters contains no photoreceptors and consti-tutes a blind spot of the eye. Several of these factors will be discussed in more detail in the subsequent sections. The physiology of the retina and the mechanisms of visual sensation are discussed in Chapter 5.

3-5 Refraction

Refraction is the change in direction of light rays as they pass from one medium to another. A common example of refraction is the effect one observes when looking down-ward into a calm pond. Objects below the surface appear nearer than they actually are. A canoe paddle, for example, will seem to have a sharp outward bend at the point where it is submerged. Refraction occurs because light travels through transparent media with different velocities, all of which are less than the velocity of light in a vacuum. The ratio of the velocity (c) of light in a vacuum to the velocity (v) in a medium is termed the index of refraction (n) of the medium:

$$n = \frac{c}{v} = \frac{\lambda}{\lambda_m}, \qquad (3\text{-}7)$$

where λ_m is the wavelength of the light in the medium. (Since the frequency of the light does not change in passing from one medium to another and since $v = \lambda\nu$ then if the velocity changes, the wavelength must also change).

The indices of refraction for the various parts of the eye and for air and water are indicated in Table 3-1.

Table 3-1
Indices of Refraction for the Eye

Material	n
cornea	1.38
aqueous humour	1.34
lens	1.41
vitreous humour	1.34
air	1.00
water	1.33

For the moment, let us consider mono-chromatic light. Because of the different vel-ocities of the light in different media, there is a discontinuity in the light ray at the interface; the light ray bends or is refracted. Consider the two examples of Figure 3-7. The change in direction that occurs is related to the indices of refraction by Snell's law,

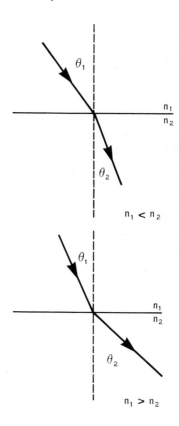

Figure 3.7
Refraction of light at a plane interface

$$n_1 \sin \theta_1 = n_2 \sin \theta_2 \qquad (3\text{-}8)$$

where the parameters are as defined in Figure 3-7. If $\theta_2 = 90°$ for the case where $n_1 > n_2$, no light penetrates into medium 2. Then, the angle of incidence is termed the critical angle and is defined as

$$\sin \theta_c = \frac{n_2}{n_1} \qquad (3\text{-}9)$$

For any angle of incidence greater than this critical angle total internal reflection takes place.

For the eye, the greatest change in index of refraction is at the air-cornea interface (Table 3-1) and thus most of the refraction takes place there. The crystalline lens has a larger index of refraction than the cornea but it is surrounded by the two humours whose indices of refraction are close to that of the lens.

Ordinary white light is made up of many colours (frequencies). Each colour will be refracted by a different amount because the index of refraction is frequency dependent. This will lead to different image positions for different colours, an effect termed chromatic aberration which will be discussed later.

3-6　Image Formation in the Eye

In the process of vision, an image is formed on the retina. Let us investigate how the elements of the eye, in particular the cornea and the lens, can lead to image formation by using the concepts of geometrical optics. The refraction of light at the cornea is essentially that which occurs at a solid spherical surface. The assumption will be made that the light rays are paraxial, that is, the angles they make with the optical axis are small (this assumption, which simplifies the geometry, is a very good one as the angles subtended at the eye are ~1°).

In Figure 3-8, O is a point object at a distance p from a cornea of radius r. C is the centre of curvature of the corneal surface. The indices of refraction outside and inside are n_1 and n_2 respectively. The incident normal ray OA, along the optical axis, is undeviated. Ray OB, incident at angle θ_1 to the normal, is refracted at an angle θ_2 and intersects ray OA at the image point I, a distance q from the cornea. From the diagram, it is evident that

$$\theta_1 = \alpha + \gamma, \qquad \gamma = \theta_2 + \beta,$$

$$\tan \alpha = \frac{h}{p+\delta}, \qquad \tan \beta = \frac{h}{q-\delta},$$

$$\tan \gamma = \frac{h}{r-\delta}.$$

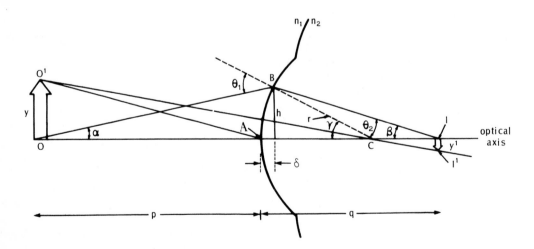

Figure 3.8
Image formation by a spherical surface

For small angles, both the sine and tangent are approximately the angles themselves in radians ($\sin\theta_1 \simeq \theta_1$, $\sin\theta_2 \simeq \theta_2$) and the small distance δ may be neglected, hence Snell's law, Equation (3-8), becomes

$$n_1(\alpha + \gamma) = n_2(\gamma - \beta) \qquad (3\text{-}10)$$

and

$$\alpha = \frac{h}{p}, \qquad \beta = \frac{h}{q}, \qquad \gamma = \frac{h}{r}. \qquad (3.11)$$

combining Equations (3-11) and (3-10) we obtain the following:

$$\frac{n_1}{p} + \frac{n_2}{q} = \frac{n_2 - n_1}{r} \qquad (3\text{-}12)$$

Given the refractive indices inside and outside the eye, we can calculate the image distance (q) for any object distance (p) in front of the cornea. It is necessary to adopt a sign convention for distances when using Equation (3-12). Since light from an object can be considered to be a flow of photons, we will consider two directions; downstream, with the flow, or upstream against the flow. The object distance will be taken as positive when the refracting surface is placed downstream from the object, and the image distance will be considered positive when the image occurs downstream from the refracting surface. The radius of curvature of the surface is considered positive when the centre of curvature lies downstream from the refracting surface. Otherwise these parameters will be taken to be negative.

For an object of finite height y, the image height y' on the retina can be determined. In Figure 3-8, the ray from the object tip through the centre of curvature C to the tip of the image is undeviated because it intersects the refracting surface at right angles. Triangles OCO' and ICI' are similar. Therefore

$$\frac{-y'}{y} = \frac{q - r}{p + r}$$

Note the additional sign convention. Distances measured up from the optical axis are positive while distances measured down from the optical axis are negative. From Equation (3-12),

$$r = \left[\frac{n_2 - n_1}{\dfrac{n_1}{p} + \dfrac{n_2}{q}}\right]$$

Thus

$$\frac{y'}{y} = -\frac{q - \left(\dfrac{n_2 - n_1}{\dfrac{n_1}{p} + \dfrac{n_2}{q}}\right)}{p + \left(\dfrac{n_2 - n_1}{\dfrac{n_1}{p} + \dfrac{n_2}{q}}\right)}$$

$$= -\left[\frac{n_1 \dfrac{q}{p} + n_2 - n_2 + n_1}{n_1 + \dfrac{n_2 p}{q} + n_2 - n_1}\right]$$

$$= -\left[\frac{n_1(1 + q/p)}{n_2(1 + p/q)}\right]$$

$$= -\left[\frac{n_1 q\,(1 + p/q)}{n_2 p\,(1 + p/q)}\right]$$

$$= -\left[\frac{n_1 q}{n_2 p}\right]$$

The ratio of the image height to the object is termed the magnification m. Therefore

$$m = -\frac{n_1 q}{n_2 p} \qquad (3\text{-}13)$$

Thus m is negative for an inverted image and positive for an erect image.

EXAMPLE

Assuming that the cornea is the only image producing element of the eye, estimate the size of the image on the retina if the object is a man (2 metres tall) standing 5 metres from the eye. The image distance (cornea to retina) is about 2.5 centimetres = 0.025 metres.

$$y' = -\frac{n_1 q}{n_2 p} y$$

$$= -\frac{1 \times 0.025 \times 2}{1.38 \times 5}$$

$$= -0.73 \text{ cm}$$

The first focal point of the refracting surface is the object position which produces an image at infinity, that is, $q = \infty$ in Equation (3-12) yielding

$$f_1 = \frac{n_1 r}{n_2 - n_1} \tag{3-14}$$

The second focal point of the refracting surface is the image position of an infinitely distant object, that is, $p = \infty$ in Equation (3-12) yielding

$$f_2 = \frac{n_2 r}{n_2 - n_1} \tag{3-15}$$

The power of an image-forming surface is defined as

$$P = \frac{n_2}{f_2} = \frac{n_2 - n_1}{r} \tag{3-16}$$

The unit of power is the diopter which is simply inverse metres. Equation (3-12) can be combined with Equation (3-16) to yield power as

$$\frac{n_1}{p} + \frac{n_2}{q} = P \tag{3-17}$$

EXAMPLE

What is the power of the cornea of the eye if the radius of curvature of the cornea is 0.008 m?

$$P = \frac{n_2}{f_2}$$

$$= \frac{n_2 - n_1}{r}$$

$$= \frac{1.38 - 1}{0.008}$$

$$= 47.5 \text{ diopters}$$

EXAMPLE

Calculate the image position produced by the cornea alone for the extremes of object position: 25 cm, the so-called near point of the eye, and infinity, the so-called far point of the eye.

For an object at the near point

$$n_1 = 1.00 \text{ (for air)}, \qquad n_2 = 1.33,$$

$$p = 25 \text{ cm}, \qquad r = 0.8 \text{ cm}$$

thus from Equation (3-12) q is given by

$$q = 3.54 \text{ cm}$$

For an object at the far point

$$n_1 = 1.00, \qquad n_2 = 1.33, \qquad p = \infty,$$

$$r = 0.8 \text{ cm}$$

thus from Equation (3-12), q is given by

$$q = 3.20 \text{ cm}$$

Note that both of these image points are beyond the retina which is 2.5 cm behind the cornea. Clearly the lens is required to complete the image-forming process.

Before proceeding to the next important refracting element in the eye, the crystalline lens, mention should be made of one of the early pioneers in the optics of vision, Herman Ludwig Ferdinand von Helmholtz (1821–1894). In addition to his famous contributions to electricity and magnetism, Helmholtz made exhaustive studies in the physiology of vision and hearing. His work in vision is incorporated in three volumes available in English translation. To the biophysicist, they make fascinating reading because he worked out in great detail most of the optical parameters for the human eye still used by opthalmologists and optometrists in modern times.

Helmholtz realized that the cornea and adjacent aqueous humour could be considered as a single refracting surface. He also assumed that the lens had a uniform refractive index when, in fact, the index increases from the surface to the centre. The value he chose for the radius of the front surface of the eye is very close to the average front corneal radius found in human adults. The placement of the lens and the lens dimensions also approximate the actual values. The optical constants of the Helmholtz schematic relaxed eye are given in Table 3-2.

Table 3-2
Optical Constants of the
Helmholtz Schematic Relaxed Eye

Optical constant	Value
Distance from the front surface of the cornea to the front of the lens	3.6 mm
Lens thickness	3.6 mm
Radii of curvature	
cornea	8.0 mm
front of the lens	10.0 mm
back of the lens	6.0 mm
Indices of refraction	
aqueous humour	1.33
lens	1.45

The lens of the eye accepts the light refracted by the cornea and provides the fine focus necessary to form a sharp retinal image. It does so mostly by varying the curvature of its front surface. This is the process of accommodation. Just how it is brought about by the muscular and nervous system is not completely understood.

The lens itself is a clear, flexible, cellular structure. The cells are long and hexagonal in cross-section and are stacked in columns roughly parallel to the light path. The layers are ordered, hence the term crystalline, in

such a way that relatively little light is scattered or absorbed. The cellular structure is illustrated in Figure 3-9(a). The lens is suspended by a complex array of ligament fibres which behave, to a first approximation, as a radial network schematically shown in Figure 3.9(b). These ligament fibres are attached to a bundle of circular muscle fibres. In the relaxed eye, the radial fibres are under tension and hold the lens at its largest radius of curvature. During accommodation, contraction of the circular muscles decreases the radius of curvature of the circular bundle and thus relaxes the radial ligaments. Because of the natural elasticity of the lens it now can increase its radius of curvature. As an object is brought nearer the eye, the image size increases thus

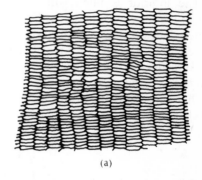

(a)

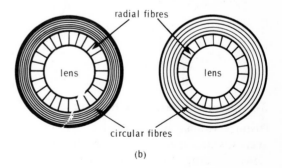

(b)

Figure 3.9
Structure of the lens of the eye (from "The Physiology of the Eye" by H. Davson, Churchill Livingstone (1972))

increasing the amount of detail that can be observed. However, in the average eye, the curvature of the lens cannot be increased to accommodate object distances closer than about 25 cm. This is called the normal near point of the eye for a person of middle age.

Refraction by the lens can be described quantitatively by using equations similar to (3-12) and calculating the refraction at both surfaces of the lens. The problem is complicated because the refractive index is not uniform throughout the lens. Nonetheless the problem has been solved with considerable accuracy.

For most calculations, a simplified eye model is assumed in which the cornea-lens doublet is replaced by a single refracting surface. The interior is filled with a medium similar to the vitreous humour with a uniform refractive index of 1.33. All distances are measured from a plane 1.67 mm behind the single refracting surface (Figure 3-10). This is sometimes referred to as the reduced eye but it has the same refracting power as the Helmholtz schematic eye.

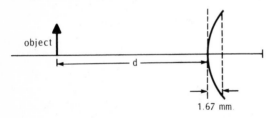

Figure 3.10
Reduced eye model

3-7 Accommodation of the Eye

Distinct vision for a normal eye is possible over a wide range of distances. If the eye is viewing nearby objects somehow it must adapt itself so that these objects at varying distances can all be brought to focus on the retina. This ability of the eye is called accommodation and is accomplished by the ability of the lens to change its shape as discussed in Section 3-6. The result is a change in the overall power of the eye.

The power of the eye for an object at an infinite distance is approximately 60 diopters. To determine the change in power effected by accommodation consider Figure 3-10. The object distance is given by

$$p = d + 1.67$$

If the image is to be formed on the retina which is 24 mm behind the front spherical surface, the image distance is

$$q = 24 - 1.67$$
$$= 22.3 \text{ mm}$$

If the object is in air then the power is given by Equation (3-17) as

$$P = \frac{1}{d + 0.00167} + \frac{1.34}{0.0223}$$

Table 3-3 shows the theoretical increase in power of the eye as the object distance d decreases. In this table ΔP corresponds to the difference $(P - P_0)$ where P_0 is the power at $d = \infty$.

Table 3-3
Change in Power of the
Eye as a Result of Accommodation

d (cm)	ΔP (diopters)
1000	0.1
100	1.0
10	9.3
5	19.4

In actual fact, the maximum increase in power as a result of accommodation ranges between 12 to 16 diopters which is obtainable only in young children. The ability to accommodate decreases with age as indicated in Figure 3-11. The result is that with advancing age the near point of the eye recedes, a phenomenon called presbyopia.

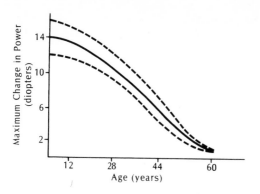

Figure 3.11
Accommodation of the eye (from "Medical Physiology and Biophysics" edited by T. C. Ruch and J. F. Fulton, W. B. Saunders Inc.)

3-8 Aberrations in Vision

The ability to perceive a clear, undistorted image of an object varies considerably from one individual to the next. Persons who see well at all distances are the exception rather than the rule and, after middle age, even this group's visual ability deteriorates. The genetic, environmental or behavioural reasons for this variability are not understood. The invention of spectacles led to the art and science of optometry and in this age, humans in all modern societies lead more active lives because of artificially improved vision.

The three most common visual anomalies are (i) hyperopia or farsightedness where the sharp image of a distant object falls behind the retina, (ii) myopia or nearsightedness where the sharp image of the distant object falls in front of the retina, (iii) astigmatism where the eye cannot focus clearly on horizontal and vertical lines simultaneously because the surface of the cornea is not spherical

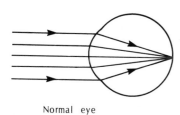

Normal eye

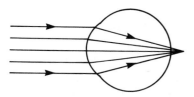

Hyperopic eye

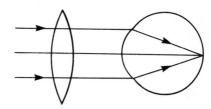

Correction for hyperopia

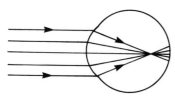

Myopic eye

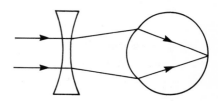

Correction for myopia

Figure 3.12
Visual aberrations and their correction

but is more sharply curved in one plane than in the other. Figure 3-12(a) illustrates image formation for the normal, hyperopic, myopic eye. Note that in the hyperopic eye the curvature of the cornea is insufficient to set the image on the back surface of the eye. In myopia, the cornea has too much curvature with respect to the shape of the eye.

In order to understand the techniques involved in visual corrections it is necessary to understand the optical properties of thin lenses.

Figure 3-13 shows cross-sections of the common lenses found in spectacles and optical instruments. Converging or positive lenses are thicker in the centre than at the edges, while diverging or negative lenses are thinner at the centre. We will assume that the lenses are perfect, that is, that the image they form will be sharp and in a single plane and all wavelengths of light are refracted in the same way.

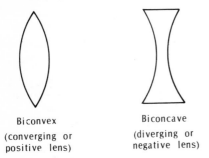

Biconvex
(converging or positive lens)

Biconcave
(diverging or negative lens)

Figure 3.13

Types of thin lenses

A thin lens is by convention one in which the entire refraction of any ray passing through it can be considered to take place at the central plane of the lens. Figure 3-14 shows light rays as they are refracted at the lens surfaces. Figure 3-15 shows rays drawn using the thin lens approximation.

The power of a thin lens equals the sum of the powers of the two refracting surfaces which compose it. Using Equation (3-16) we obtain, for a lens sitting in air

$$P = P_1 + P_2 \tag{3-18}$$

$$= \frac{n_2 - 1}{r_1} + \frac{1 - n_2}{r_2}$$

$$= (n_2 - 1)\left(\frac{1}{r_1} - \frac{1}{r_2}\right) \tag{3-19}$$

where r_1 and r_2 are the radii of curvature of the two surfaces of the lens material with refractive index n_2. This is the lensmakers equation. For a lens in a different medium with refractive index n_1,

$$P = (n_2 - n_1)\left(\frac{1}{r_1} - \frac{1}{r_2}\right) \tag{3-20}$$

The equation becomes even more complicated when the media on each side of the lens differ.

If very close attention is paid to the sign convention stated following Equation (3-12) it is possible to combine Equation (3-20) with Equation (3-12) and obtain

$$P = \frac{1}{f} = \frac{1}{p} + \frac{1}{q} \tag{3-21}$$

where f is the focal length of the lens. In arriving at Equation (3-21) it must be realized that the image produced by the first refracting surface is the object for the second. One can also show that

$$m = -\frac{q}{p} \tag{3-22}$$

Figure 3-14 illustrates the refraction of light by a converging and diverging thin lens. It shows the definition of the focal length, f, in Equation (3-21). The focal length is positive for a converging lens and negative for a diverging lens. Hence these lens are often termed positive and negative respectively.

Geometrical ray diagrams can also be used

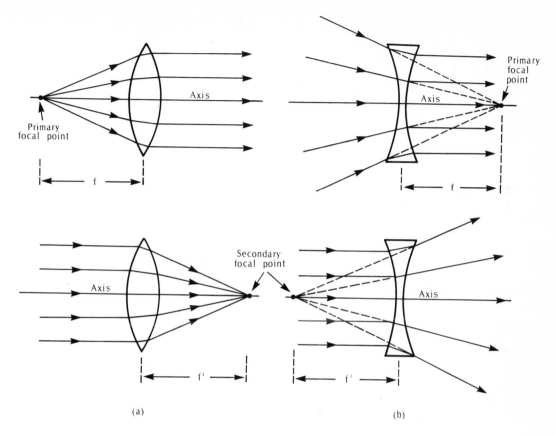

Figure 3.14
Refraction of light by a converging (a) and diverging (b) thin lens

to determine image positions as shown in Figure 3-15 for the case of a positive and a negative thin lens. Some of the important points to be considered in making such diagrams are as follows. First for the positive lens.

(a) the central ray is undeviated,
(b) an object ray passing through the front focal point is rendered parallel in the image light,
(c) a parallel ray from the object passes through the back focal point of the lens,
(d) when the object is further from the lens than the focal point the image is real and

inverted. A real image is visible on a screen placed at the image position. When the object is placed inside the focal length, the image is virtual (see below), erect and on the same side of the lens as the object.

For the negative lens.

(a) the central ray is undeviated,
(b) an object ray aimed at the back focal point is made parallel in the image light,
(c) a parallel ray from the object is diverted so that its backward projection goes through the front focal point,
(d) the image is in front of the lens, is erect

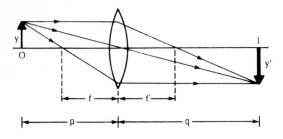

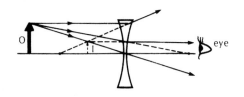

Figure 3.15
Geometrical method for finding the image position for thin lenses

and is virtual. All object positions for negative lenses give rise to virtual images. Light does not come to a focus at virtual images, it just appears to do so and thus they cannot be observed on a screen placed at the image position. However they can be seen by the eye. You see a virtual image of yourself every morning in the mirror.

EXAMPLE

An object 0.026 m high is placed 0.5 m in front of a lens having a power of 3 diopters. Find the size and position of the image by both geometric and algebraic means.

(a) Geometric method

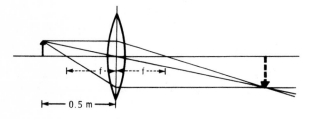

(b) Algebraic method

$$P = \frac{1}{p} + \frac{1}{q}$$

$$3 = \frac{1}{0.5} + \frac{1}{q}$$

$$\frac{1}{q} = 1 \quad \text{or} \quad q = 1 \text{ m}$$

$$m = -\frac{q}{p} = -\frac{1}{0.5} = -2$$

It is also a simple step to combine the action of lenses as one does, for example, in constructing a projector, a microscope, or a telescope. In such cases the image produced by the first lens is the object for the second. Thus the power of the combined lens system is just the sum of the powers of the component lenses. However the magnification rendered by the combined lens system is the product of the separate lens magnifications.

$$M_{\text{system}} = m_1 m_2 \qquad (3\text{-}23)$$

EXAMPLE

Find the magnification and image position of an object placed 0.15 m from the first lens of the two lens system shown below. The focal lengths of the two lenses are 0.13 m for the first lens and 0.08 m for the second lens. The two lenses are separated by a distance of 0.45 m. For first lens:

$$\frac{1}{f} = \frac{1}{p} + \frac{1}{q},$$

$$\frac{1}{0.13} = \frac{1}{0.15} + \frac{1}{q'}$$

$$q' = 1 \text{ m}$$

The object position at q' yields the image distance p' for the second lens. Since q' was downstream from the second lens p' is negative.

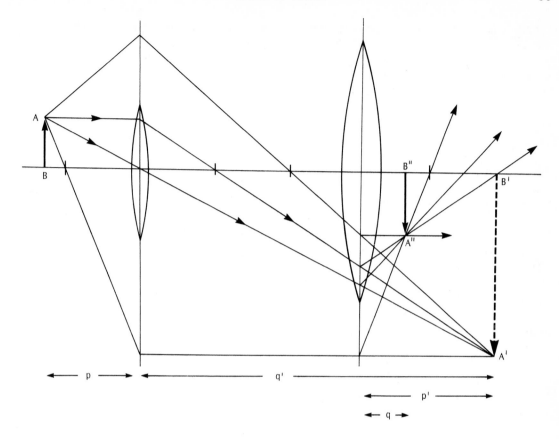

Thus

$$p' = -(1.00 - 0.45) = -0.55 \text{ m}$$

$$\frac{1}{0.08} = \frac{1}{-0.55} + \frac{1}{q}$$

$$12.5 = -1.82 + \frac{1}{q}$$

$$q = \frac{1}{14.32} = 0.07 \text{ m}$$

The corresponding magnification of this system of lenses is

$$M = m_1 m_2 = \left(-\frac{q'}{p}\right)\left(-\frac{q}{p'}\right)$$

$$= \left(-\frac{1}{0.15}\right)\left(-\frac{0.07}{(-0.55)}\right)$$

$$= -0.85$$

The real image is inverted and 0.85 times the size of the object.

Let us now investigate how thin lenses can be used to correct for hyperopia, myopia, and astigmatism. Figure 3-12(b) shows the effect of the lens required to correct hyperopia and myopia. The following examples demonstrate how to determine the parameters of the necessary lenses.

EXAMPLE: THE HYPEROPIC EYE

A person with this defect would live in a fuzzy world (except when using corrective lenses) because his (relaxed) eye cannot focus a sharp image on the retina for any object distance. Such a condition results when the eye lens is removed in cataract surgery. It is also present in many healthy eyes. The use of the proper convex lens in front of the eye brings the rays to a sharp focus on the retina (Figure 3-12(b)). Sometimes lenses of different powers are required for distant, intermediate and close-up vision. These are often compounded into bifocal and trifocal spectacles.

Suppose a person can by accommodation see clearly an object 1 m in front of the eye. What power lens would be required for this person to be able to see clearly an object 0.25 m in front of the eye?

Clearly a converging lens is necessary. The diagram shows the effect. The object at a distance p of 0.25 m forms a virtual image at a distance $q = -1$ m.

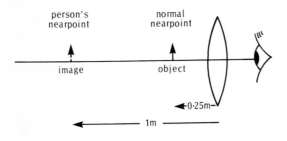

Using Equation (13-21) the power of the lens is

$$P = \frac{1}{0.25} + \frac{1}{-1}$$

$$= 3 \text{ diopters}$$

EXAMPLE: THE MYOPIC EYE

The myopic eye focuses parallel light in front of the retina and hence cannot see well at a distance. The corrective lens must render distant light less convergent and produce a virtual image at the normal far point.

The far point of a myopic eye is 5 m. What power is required in order for the eye to see clearly an object at infinity?

Clearly a diverging lens is required. The diagram shows the effect. The object at infinity forms a virtual image at a distance $q = -5$ m.

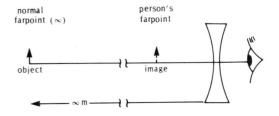

Using Equation (3-21) the power of the lens is

$$P = \frac{1}{\infty} + \frac{1}{-5}$$

$$= -0.2 \text{ diopters}$$

EXAMPLE: THE ASTIGMATIC EYE

People with this defect have distorted vision because of the non-uniform curvature of the cornea or lens surfaces. In the simplest case the refracting elements have two focal lengths. One image axis will be in focus while the other perpendicular axis is out of focus. The correction in this simple case is a cylindrical lens, that is, one with no curvature in one plane and sufficient curvature in the other to bring the two image axes into coincidence.

Before we leave the topic of visual aberrations something should be said about the problem of chromatic aberration of the lens of the eye. The index of refraction of a material varies with the wavelength of the light so that all colours will not be focussed at the same point by an optical system. In optical systems double lenses which compensate for various wavelengths can be used to eliminate chromatic aberrations. But the eye contains one lens made of one material so the eye has a high degree of chromatic aberration. However the effect doesn't bother us very much for the following reasons. The effect is moderate in the red and green, but becomes very pronounced in the ultraviolet. However, the lens of the eye is relatively opaque in the ultraviolet. Also, in dim light, we see by rod vision which is most sensitive in the blue-green. This is a coarse grained type of vision. In bright light, the eye shifts to cone, high acuity vision, whose sensitivity is better in the red. Here the chromatic aberration is least. Nature is very clever.

3-9 Binocular Vision

Humans have two eyes, both of which look in the forward direction. The fields of view of the two overlap for about 170°. This arrangement allows for very accurate determination of distance and relative position. Thus humans have binocular vision. On the other hand, some animals, such as chameleons, have panoramic vision, that is, their eyes look in different directions with no overlap of the fields of view.

Because the human eyes are separated, a slightly different image is viewed by each eye—a phenomenon called binocular parallax—thus, the two images must be meshed by the visual system. This gives rise to depth perception. For short distances, we use binocular parallax to judge distance. For very distant objects, the two eyes see practically the same view and binocular parallax effects no longer play a role. Then distance is judged by the size of familiar objects, the amount of light scattered by intervening air, and the curvature of the horizon.

To demonstrate to yourself the effect of binocular parallax, hold out your arms at the side with your index fingers pointed out. Swing your arms around in front to bring your index fingers to meet at the same point. Use one eye, then repeat the experiment using both eyes.

3-10 Resolution of the Eye

Up to this point we have been working under the assumption that light could be treated by the ray concept. But the laws of geometrical optics are valid only if the dimensions of the physical objects that the light encounters are much larger than the wavelength of the light. However if light passes through small apertures some diffraction effects (bending of the light ray by the obstacle) will be seen.

Consider the image on a screen of a small pin-hole which is uniformly illuminated with parallel light. The image should be a geometrical point and it will be if the aperture (pinhole) is large. However, if the aperture is small, diffraction will occur leading to a blurring of the image. Light from a point source diffracted by a circular opening is focused then, not as a point but as a disc of finite radius surrounded by bright and dark rings. Two point sources can be resolved by an optical system if the corresponding diffraction patterns are sufficiently small or sufficiently separated to be distinguishable. The usual criterion for resolvability is that stated by Lord Rayleigh, "Two equally bright point sources can just be resolved by an optical system if the central maximum of the diffraction pattern of one source coincides with the first minimum of the other" (Figure 3-16). Thus the distance between the centres of the diffraction patterns equals the radius of either central disc. For a circular aperture, this distance in angular measure (radians) is given by

$$\alpha_2 = 1.22 \frac{\lambda_m}{a} \qquad (3-24)$$

where λ_m is the wavelength of the light in the eye (λ/n_2), a is the diameter of the aperture and α_2 is the angular separation (Figure 3-16).

The pupil of the eye is a variable aperture

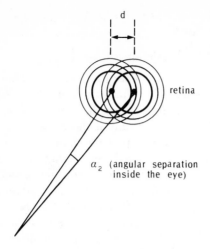

Figure 3.16
Rayleigh resolution criterion

and diffraction effects can be produced. The minimum diameter of the pupil is about 2 millimetres and the eye is most sensitive to light of a wavelength of about 500 nm. If we assume that all refraction takes place at the cornea (Figure 3-17) then the resolution limit of the eye is

$$\alpha_2 = \frac{1.22(500 \times 10^{-9})}{1.33 \times 2 \times 10^{-3}}$$

$$= 2.3 \times 10^{-4} \text{ radians}$$

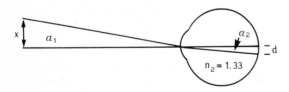

Figure 3.17
Inside and outside angles of resolution of the eye

This angular separation can be converted to a linear separation, d, at the retina using the relationship;

$$\alpha_2 = \frac{d}{0.025}$$

in which 0.025 m is the corneal-retinal distance. Thus

$$d = 2.3 \times 10^{-4} \times 0.025$$
$$= 5.8 \times 10^{-6} \text{ m}$$
$$= 5.8 \ \mu\text{m}$$

The angle of resolution outside the eye, α_1, can be evaluated using Snell's law.

For small angles $\sin \alpha_1 \simeq \alpha_1$ and $\sin \alpha_2 \simeq \alpha_2$. Therefore,

$$n_1 \alpha_1 = n_2 \alpha_2$$
$$\alpha_1 = 1.33 \times 2.3 \times 10^{-4}$$
$$= 3.1 \times 10^{-4} \text{ radians}$$

The average cone separation in the fovea is about 2 μm so that the minimum distance of 5.8 μm calculated above would span approximately three cones. This seems to be necessary to effect a visual response. Two active cones must be separated by a "cold" cone in order to create the sensation of an image point. Note the inverse dependence on pupil diameter. Wider pupils should give better resolution. How does this affect your prejudices about the use of sunglasses? The reduced illumination would lead to wider pupils and the depth of field would tend to narrow as in Figure 3-18. Also any astigmatism is more pronounced. Most people see most comfortably under conditions of good illumination.

The short calculation above also demonstrates that the angular separation of objects must be about 3×10^{-4} radians. We can find the required separation (x) of two objects placed at the nearpoint of the normal eye. Since

$$\alpha_1 = \frac{x}{\text{(distance to nearpoint)}}$$

Then
$$x = (0.25)(3 \times 10^{-4})$$
$$= 7.5 \times 10^{-5} \text{ m or } 75 \ \mu\text{m}$$

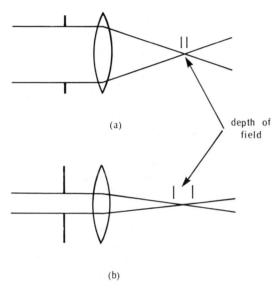

(a)

depth of field

(b)

Figure 3.18
(a) Large pupil diameter – small depth of field
(b) Small pupil diameter – large depth of field

This is approximately, then, the nearest separation at which we can still see distinct objects. Obviously, with the aid of the appropriate lens system, such as a microscope, we can do much better. The resolution of a microscope is diffraction limited, just as the eye, but gains a significant advantage because an object can be placed very close to its objective lens. The smallest resolvable separation x_m for an ordinary microscope is given by

$$x_m = \frac{\lambda}{2n \sin \theta} \qquad (3\text{-}25)$$

where λ is the wavelength of the light used and θ is half the viewing angle of the objective as shown in Figure 3-19. The quantity $n \sin \theta$ is called the *numerical aperture* of the objective. The better microscope objectives have numerical aperture values of about 0.5. Thus with the aid of a microscope and 500 nanometre light, we can see objects distinctly at separations of only

$$x_m = \frac{500 \times 10^{-9}}{2 \times 0.5} = 0.5 \ \mu\text{m}$$

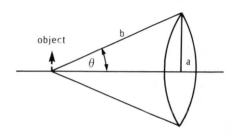

Figure 3.19
The numerical aperture: n sin θ=a/b

This value can be slightly improved by placing oil between the object and the objective, the so called oil immersion microscope. In this type of microscope the resolution is increased because of the relatively large value of the refractive index of oil.

Problems

1. If red light has a wavelength of 620 nm in a vacuum, what is the wavelength, frequency, and velocity of this light in the vitreous humour of the eye? ($n = 1.33$)
2. An object under water appears closer to the surface than it really is. Why?

3. What is the velocity of light in medium 2?

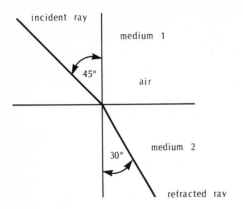

4. The minimum diameter of the pupil of the human eye is about 2 mm and the maximum diameter is about 8 mm. How much more light energy per unit time enters the eye with the pupil fully dilated than when the pupil is at its minimum size?

5. The diameter of the image of the sun on the retina is only about 160 μm. Because of the sun's brightness, we will normally blink so that the image lasts only about 0.2 s. The intensity of the light at the retina for such an image is about 10^4 W m^{-2}. If we imagine that the light of the sun is of one wavelength only, say 500 nm, how many photons does this short glimpse of the sun involve?

6. A thin lens has radii of curvature $r_1 = +2$ cm and $r_2 = +1$ cm. The glass of which it is made has refractive index 1.5.
    (a) What is its power in air?
    (b) What is its power in water? ($n = 1.33$)
    (c) Now suppose that it is placed in a wall on one side of which is air and on the other water. (Surface with radius r_2 in water). What is the power in this case?
    (d) In cases (a), (b), and (c), calculate where the two focal points are.
    (e) In cases (b), and (c) find out where the image of an object placed 10 cm to the left of the lens is formed. What power would the lens in case (b) have to have in order that the image distance be the same as in case (c)?

7. A person with hyperopia has a near point of 1.00 m. A lens of power +2.00 d is placed directly in front of the eye. What is the near point of the eye-plus-lens system?

8. The far point of a myopic eye is 2.00 m. What is the power of the spectacle lens which will correct the myopia? What is the focal length of this lens?

9. The reduced eye (i.e. the simplified model of the human eye) is useful in calculating where images are formed and what sorts of corrections must be made to improve a faulty lens system in the eye.
    (a) Suppose that the only thing which happens when the eye accommodates is that the power of the lens changes. What does it have to change to in order that an object 20 cm from the front surface of the eye is in focus on the retina? What increase in power above the power of the reduced eye must accommodation therefore provide?
    (b) The fovea of the human eye is approximately 0.3 mm in diameter. At what distance would a circle of diameter 10 cm have to be so that the image covers the fovea?

10. Two small lights are placed 10 metres from your head in a darkened room. Estimating any data you require, calculate the minimum separation of these lights such that you still recognize them as being two.

11. Two stars make an angle of 10^{-3} radians at the eye of an observer. If a person who can just distinguish two spots, 0.5 cm apart, at a distance of 10 m views the stars, can he distinguish them as two distinct objects?

12. Two lenses having equal focal lengths of 0.5 m are separated by 0.8 m. If a 0.3 m high object is placed 4 m in front of this lens system find the position and height of the image.

13. Wally Whoop has been a bird-watcher for many years. However, he has always been frustrated in his desire to observe the nesting habits of his favourite birds, cranes. When he got too close, they flew away. Too poor to buy binoculars or a telescope, Wally had just enough money to buy two lenses and a jar of rubber cement. The focal length of one lens, the objective, was 0.30 m. This lens was placed at one end of a cardboard tube from a roll of wax paper. Wally didn't know the focal length of the other lens, the eye lens, but he kept repositioning it until he obtained a system which would magnify objects 50 metres away by a factor of ten.

What is:

(a) the separation of the lenses

(b) the focal length of the eye lens?

(Hint: The image of the lens system should be roughly at infinity (assume 50 m from the objective). Note also that for distant objects the image falls on the focal point.)

Chapter 4

Absorption and Emission of Light by Molecules

4-1 Introduction

Many biological molecules are particularly important because of their ability to convert the energy of an electromagnetic wave into chemical energy. Well known examples of these molecules are chlorophyll in plants and the various pigments involved in vision. An enormous amount of scientific effort has gone into studies of light absorption by molecules resulting in the development of many new disciplines such as molecular spectroscopy, photochemistry, photobiology and photophysiology. The theories which have evolved are certainly not simple ones. However, light absorption is so vital for life itself that everyone should know something about the process.

To begin we need to remember that ultraviolet (UV) and visible light waves constitute only a small fraction of the electromagnetic spectrum. A photon of light at the gamma ray end of the spectrum has energy of about 10^{-13} joule, while a photon near the radio end of the spectrum has an energy of only 10^{-28} joule. Thus, there is an enormous difference between the energies of photons from different parts of the spectrum. We should also remember that the energies involved in most chemical reactions are of the order of 30–100 Kcal mol^{-1} or 2×10^{-19}–7×10^{-19} joule per molecule. Photons in the UV and visible part of the spectrum have energies which are right in this range. It should, therefore, not be very surprising to learn that these photons are involved in photochemical and photobiological reactions. Since chemical bonds are primarily interactions between electrons, it is apparent that the initial effect of the absorption of a photon is the redistribution of electrons in the molecule.

Thus, photons in the UV-visible region have a special ability to initiate chemical reactions when they are absorbed. By contrast, photons in the x-ray or γ-ray region are so energetic that they blast the molecule apart and, hence, such photons are said to be ionizing radiation. Studies of the interaction of ionizing radiation and matter have led to development of such disciplines as radiation physics, radiochemistry and radiation biology, a subject area which will be discussed later. At the other end of the spectrum, photons of infrared and microwave radiation have very little energy and in relative terms only "tickle" the molecule. These photons induce changes in the ways in which molecules vibrate and rotate. Detailed infrared absorption investigations have led to an extensive knowledge of properties and structure of molecules. Thus, while the basic physical characteristics of all photons are the same, the particular effect a photon has on a molecule depends solely upon the photon energy.

For light to be absorbed by a molecule, it is obvious that there must be some kind of interaction between the light wave and the particles which constitute the molecule. We have already devoted considerable space to a discussion of the properties of electromagnetic waves. It remains for us to discuss the properties of the particles which make up molecules, electrons especially, and discuss their role in the light absorption process. These topics serve as stepping stones toward a detailed discussion of the visual processs.

4-2 Wave Properties of Particles

We have just pointed out that electrons in atoms and molecules are inherently involved in the absorption and emission of light. In order to pursue these phenomena in any detail we must know considerably more about the behaviour of electrons in these species. Here, however, we encounter a major stumbling block in the way we like to visualize things. The problem is that we like to explain various objects and events in terms of the macroscopic things we experience every day. However, there is no particular reason to believe that microscopic phenomena can be explained in this way. In fact, our macroscopically oriented judgments and intuitions completely fail to provide a description of microscopic particles and events. It is this unexpected realization which makes physics fascinating to some people and incomprehensible to others. For example, in the case of electrons in atoms, a person applying his macroscopically based knowledge would feel that he could predict and explain every phenomenon if he could simultaneously know the exact positions, velocities, momenta, etc., of all the components of an atom. He would be staggered to learn that there is a principle, experimentally proven in the twenties, which expressly forbids us knowing the exact values of these parameters at any time. This principle, known as the Heisenberg Uncertainty Principle may be written as

$$\Delta v\, \Delta x \simeq h/m_o$$

where Δv is the uncertainty in velocity of a particle, Δx is the uncertainty in the position of the particle, h is Planck's constant and m_o is the mass of the particle in question.

Let us apply this idea to large and small particles respectively. Consider first of all a flying Canada goose weighing 4.5 kilograms. The product of the uncertainties in position and velocity is about

$$\frac{6.63 \times 10^{-34}}{4.5} \simeq 10^{-34}\, m^2\, s^{-1}$$

This is such a small number that in a macroscopic sense there is virtually no uncertainty involved. Consider next, however, an electron in an atom. In this case, the product of the uncertainties is

$$\frac{6.63 \times 10^{-34}}{9.11 \times 10^{-31}} \simeq 10^{-3}\, m^2\, s^{-1}$$

which is huge number relative to the value obtained for the Canada goose. The large value of this uncertainty becomes even more apparent if, for example, ʰe position of the electron is measured in an experiment to about ± 0.1 nm which is the approximate diameter of a hydrogen atom. The corresponding uncertainty in the velocity of the electron would then be

$$\Delta v = \pm \frac{10^{-3}}{10^{-10}} = \pm 10^{7}\, m\, s^{-1}$$

With this large an uncertainty it becomes ridiculous to try to evaluate a velocity of the electron at all. On the other hand if the experiment is set up in a different way such that one can measure the velocity to $\pm 10\, m\, s^{-1}$ then the corresponding uncertainty in position would be about a million times the radius of a hydrogen atom. You can see, therefore, that it is hopeless to try to analyse the electron in an atom with familiar macroscopic concepts. If one tries to improve the experiment by measuring one quantity more accurately, the uncertainty in the other gets worse. One is, therefore, royally wrecked.

The problem we are experiencing here' was a familiar and puzzling one to scientists in the twenties. Their debates on the subject were invariably emotional. Slightly earlier, however, an unknown French graduate student Louis de Broglie made a hypothesis which eventually helped clear up the dilemma. He suggested that we scrap our attempts to describe small objects such as the electron in purely particle terms, but instead describe them in terms of both waves and particles. Further, he predicted that the wavelength, λ, of a particle could be defined by the equation,

$$\lambda = h/p \qquad (4\text{-}1)$$

where h is Planck's constant and $p = mv$ is the momentum of the particle. This hypothesis was so far out that few scientists even seriously discussed it until 1926 when two American physicists, Davisson and Germer, proved experimentally that, indeed, the electron does have wave nature. They found that a beam of electrons could be diffracted by a crystal just as electromagnetic waves (x-rays) are diffracted by a crystal. To the chagrin of many they proved that the de Broglie equation was correct. In particle waves, the quantity which is oscillating is related to a quantity called the probability density, P_x. For now, we'll operationally define probability density such that if the probability density is large at some point, then the likelihood of finding a particle at that point is large. If the probability density at some point is low, then the likelihood of finding a particle there is low.

Since particle waves are used to describe electrons in atoms and molecules before and after light absorption, it is fairly easy to confuse the properties of particle and electromagnetic waves. Table 4-1 lists the properties of both waves and hopefully will remove some of the confusion.

4-3 Absorption of Ultraviolet-Visible Light by Molecules

In the introduction to this chapter, it was stated that the initial event following the absorption of a UV or visible photon was a restructuring or redistribution of the electrons in the molecule. This implies that if the process is to be understood we must have some knowledge of the distribution of electrons before and after the absorption event. We've already seen that it is hopeless to try to determine exact positions, velocities, and momenta of electrons in molecules. The proper way of attacking the problem involves a consideration of the wave properties of the electron. Thus, in this section we will be using standing particle waves in order to obtain

Table 4-1
Comparison of Electromagnetic and Particle Waves

Electromagnetic Waves	Particle Waves
magnetic and electric fields oscillate	P_x oscillates
travel in packets called photons	particles travel singly or in beams
produced by accelerating a charge	particle beams can be produced by accelerators, reactors or potential gradients
energy of a photon $E = h\nu$	energy of a particle $E = \frac{1}{2}mv^2$
travels with speed of light c	can travel at any velocity $(<c)$
wavelength $\lambda = c/\nu = hc/E$	wavelength $\lambda = h/p = h/mv$

probability densities and hence the distribution of electrons. We will find that we gain further benefits by this approach in that we will be able to calculate the energies of the electrons in the molecule and the approximate wavelengths of light the molecule will absorb.

If you're of the opinion that this approach, which is called the quantum mechanical approach, is complicated and difficult, you're right. Still, a large number of chemists, molecular biologists and biophysicists use the theory in attempts to explain various biological phenomena such as mutations, macromolecular structure, and enzyme reactions

in addition to the photobiological processes we are about to consider. Actually, we are quite fortunate in that absorption of UV and visible light by many biological molecules can be substantially simplified by a few approximations.

The first approximation is that molecules composed of carbon, hydrogen, oxygen, and nitrogen can essentially be considered to contain two groups of electrons, namely the σ electrons and π electrons which arise from σ bonds and π bonds respectively. Figure 4-1 shows electron clouds corresponding to the σ and π bonds of ethylene.

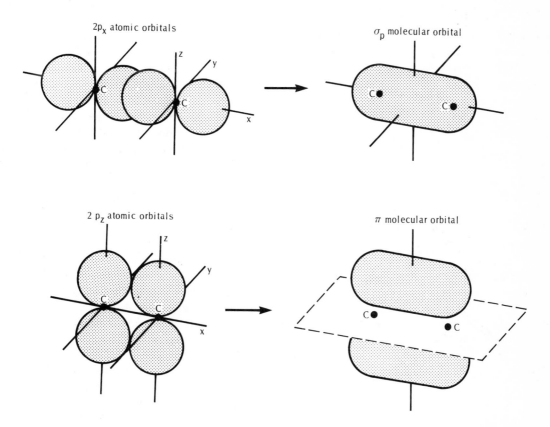

Figure 4.1

π and σ electrons in ethylene

The words "electron cloud" refer to probability density clouds which we have not yet discussed. For now, just remember that where the electron cloud is most dense, the probability of finding the electron is highest. The electron cloud of π electrons has been shaded in. Note that each carbon atom (or oxygen or nitrogen) contributes 1 π electron to the cloud. For such molecules to absorb UV or visible light, they just have a series or chain of conjugated double bonds such as in octene, an organic molecule, or cytosine, part of DNA.

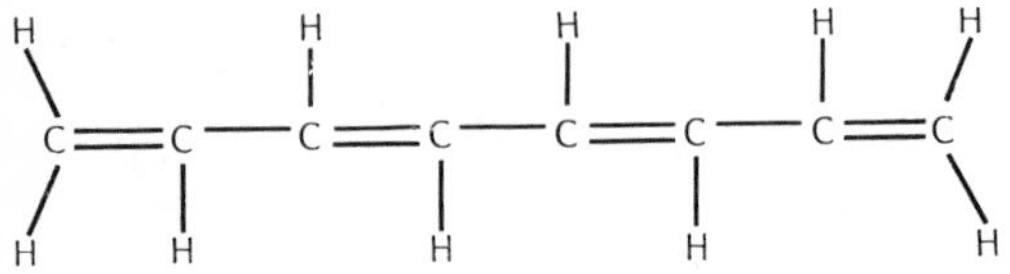

Figure 4.2
Octene

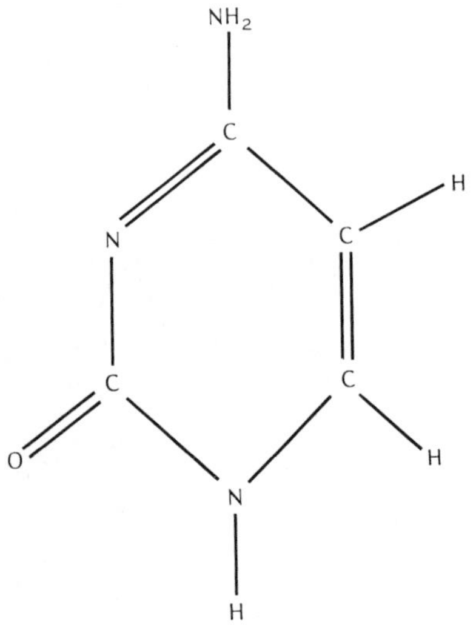

Figure 4.3
Cytosine

In these molecules, the π electrons are free to roam anywhere in the conjugated double bond chain. Since the π electrons are not as tightly bound as σ electrons, they undergo excitation comparatively easily. This brings us to our second approximation: π electrons alone are involved in the absorption of UV or visible light. Thus, it is assumed that absorption of UV or visible light brings about a restructuring or redistribution of only the π electrons in the molecule; σ electrons will be completely ignored.

First we will try to find the distribution and energy of π electrons before the photon arrives i.e. when the molecule is in the ground electronic state. The first example we will choose is vitamin A, a molecule important in vision.

All of the ten π electrons in this vitamin A molecule are free to wander throughout the long chain of conjugated double bonds. Since the particle wave associated with the electron is related to the probability density of the electron, it too must exist along this conjugated double bond sequence. We will now try to find a particle wave function, given the symbol ψ, which corresponds to each of the ten π electrons in the conjugated double bond sequence.

Let's first string out these double bonds in a straight line to simplify the calculation. We can consider that this string of carbons corresponds to a 'box' in which the x axis (horizontal axis) runs along the length ℓ of the molecule as in Figure 4-5.

The length of the box is roughly the length of nine carbon–carbon bonds which is $9 \times 0.15 = 1.35$ nm.

Let's add one π electron of the ten π electrons to this box. Once we've added it, there must be a finite probability of finding the particle at some point inside the box, but the probability of finding the electron outside the box must be zero. We need a ψ, therefore,

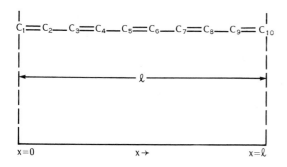

Figure 4.4

Vitamin A

$C_1{=}C_2{-}C_3{=}C_4{-}C_5{=}C_6{-}C_7{=}C_8{-}C_9{=}C_{10}$

ℓ

$x=0$ $x\rightarrow$ $x=\ell$

Figure 4.5

"Box" corresponding to vitamin A

which has a value inside the box, but is zero everywhere else. There are many functions we could choose and often it's difficult to know which is the best one. We'll choose a simple one though;

$$\psi = b \sin \frac{\pi x}{\ell} \qquad (4\text{-}2)$$

where x can have values of 0 to ℓ only, (i.e. $0 < x < \ell$) and b is a constant. If

$$x = 0$$

then

$$\sin 0 = 0$$

and if

$$x = \ell$$

then

$$\sin \frac{\pi \ell}{\ell} = \sin \pi$$

$$= 0$$

and the function, when superimposed on the box, looks like the one shown in Figure 4-6. Note that a cosine function will not fit the boundary conditions and it cannot be considered as a possible wave function.

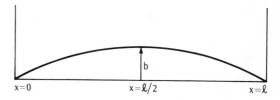

$x=0$ $x=\ell/2$ $x=\ell$

Figure 4.6

Diagram of $\psi = b \sin \frac{\pi x}{\ell}$ in the "Box"

We have found a wave function ψ which meets our requirement. According to the ideas presented earlier, this function describes the shape and density of the electron cloud. At the centre of the box (which corresponds to the mid-point of the bond between C_5 and

C_6) the cloud is most dense. The probability density (P_x) of finding the electron is greatest in this region. If one wants to calculate this probability density at some point, one would use the equation

$$P_x = \psi_x^2 \qquad (4\text{-}3)$$

where ψ_x is the value of ψ at the point, x.

There are many rigorous reasons why P_x is equated to ψ_x^2 rather than ψ_x, but for now you may consider that it is simply due to the fact that one cannot have a negative probability and ψ_x^2 is always positive (even if ψ_x is negative).

We now need to devote a little bit of our time to the constant b which appeared in front of the sine. First let us superimpose $P_x = \psi_x^2$ inside the box, Figure 4-7.

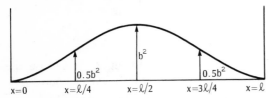

Figure 4.7

Diagram of $\psi^2 = b^2 (\sin^2 \frac{\pi x}{\ell})$ in the "Box"

At $x = \ell/4$ the probability density P_x of finding the electron is $0.5b^2$. This corresponds to a point on the vitamin A molecule between C_3 and C_4. At each point along x, there is a corresponding probability density P_x. However, we know that we put one electron in the box and if we sum up all the probability densities over the whole box we must end up with the total probability of finding the electron in the box. Since we put it in there and it cannot be anywhere else, that probability is 1. Mathematically, this summation corresponds to an integration of the probability density over the box. Thus,

$$b^2 \int_{x=0}^{x=\ell} \left(\sin \frac{\pi x}{\ell} \right)^2 dx = 1$$

This is a tricky integration, but it can be done and one would find that,

$$b = \sqrt{\frac{2}{\ell}}$$

The wave function is then written

$$\psi = \sqrt{\frac{2}{\ell}} \sin \frac{\pi x}{\ell}$$

This equation for ψ is one of a whole family of functions of the form:

$$\psi_n = \sqrt{\frac{2}{\ell}} \sin \frac{n \pi x}{\ell}, \qquad n = 1, 2, 3, \ldots \quad (4\text{-}4)$$

each of which are wave functions which satisfy the requirements which were specified. Wave functions ψ_2 and ψ_3 are shown in Figure 4-8 and Figure 4-9, respectively.

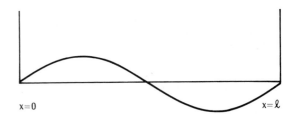

Figure 4.8

Diagram of $\psi_2 = \sqrt{\frac{2}{\ell}} \sin \frac{2\pi x}{\ell}$ in the "Box"

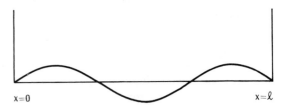

Figure 4.9

Diagram of $\psi_3 = \sqrt{\frac{2}{\ell}} \sin \frac{3\pi x}{\ell}$ in the "Box"

Inspection of these figures indicates that the wave functions, for different values of n, are similar to the various possible standing waves on a string (see Appendix 3). Indeed, the wave functions described by Equation (4-4) are standing particle waves of the π electrons in the molecule. The time dependent part of the standing wave (the $\cos \omega t$ part) has been ignored because our interest is centred on the spatial location of the electrons. The number $n = 1, 2, 3, \ldots$ is called a quantum number and specifies the number of electron half-wavelengths $(\lambda_e/2)$ which must be placed in the box.

The units of the probability density, P_x, are apparent by squaring Equation (4-4); probability per unit length. In order to obtain a probability, the probability density must be multiplied by a length. The probability of finding the electron in a region of length Δx is

$$P_{\Delta x} = P_x(\Delta x) \qquad (4-5)$$

If P_x is constant over the region Δx then this equation is satisfactory. However, if P_x is changing across the region then an integration is necessary:

$$P_{\Delta x} = \int_x^{x+\Delta x} P_x \, dx \qquad (4-6)$$

EXAMPLE

(a) Find the probability density of an $n = 3$ electron at the point indicated by the arrow in the wave function diagrammed below.

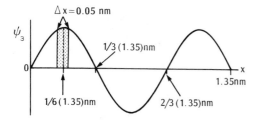

(b) Find the probability of finding the electron in the shaded region.

(c) What would be the probability of finding the electron between $x = 0$ and $x = 1.35/3$?

(a) $P_x = \psi_{3x}^2 = \dfrac{2}{1.35} \sin^2 \left[\dfrac{\left(3\pi \dfrac{1.35}{6}\right)}{1.35} \right]$ from

Equations (4-3) and (4-4)

$P_x = \dfrac{2}{1.35}(1) = 1.5 \text{ nm}^{-1}$.

(b) Equation (4-5) can be used if P_x is assumed to be constant along Δx. (From the diagram this is not too bad an approximation if Δx is small.) Thus,

$$P_{\Delta x} = 1.5 \times 0.05 = 0.075$$

Therefore the probability of finding the $n = 3$ electron in the shaded region is 0.075 or 7.5 percent of the time.

(c) If the function diagrammed above was squared, one third of the area under the resulting curve would be between $x = 0$ and $x = 1.35/3$. Therefore the probability of finding the electron in this region is $\frac{1}{3}$ or 33.3 percent of the time.

Let us determine the energy of a π electron having $n = 1$. Since $n = 1$ we would have to fit one half-wavelength inside the box as in Figure 4-7. Since the wavelength of ψ is the wavelength of the electron then

$$\lambda_e/2 = \ell$$

or

$$\dfrac{h}{2m_e v} = \ell$$

because $\lambda_e = h/m_e v$ (the de Broglie equation). Squaring the above leads to

$$\dfrac{h^2}{4m_e^2 v^2} = \ell^2$$

But the kinetic energy of a particle such as the electron is $E_k = \frac{1}{2}m_e v^2 = m_e^2 v^2/2m_e$. This kinetic energy is very near the total energy of the π electron, E. Thus,

$$m_e^2 v^2 = 2 m_e E$$

$$\frac{h^2}{8 m_e E} = \ell^2$$

and

$$E = \frac{h^2}{8 m_e \ell^2}$$

The energy of this π electron in vitamin A is then

$$E_1 = \frac{(6.63 \times 10^{-34})^2}{8(9.11 \times 10^{-31})(13.5 \times 10^{-10})^2}$$

$$= 3.31 \times 10^{-20} \text{ joules}$$

If $n = 2$, we would find by the same procedure as before that

$$E_2 = \frac{4h^2}{8 m_e \ell^2} = 1.33 \times 10^{-19} \text{ joules}$$

and generally that

$$E_n = \frac{n^2 h^2}{8 m_e \ell^2}, \qquad n = 1, 2, 3, \cdots \quad (4\text{-}7)$$

Be sure that you do not confuse the letter n which we have used in these π electron calculations with the principal quantum number n which you might have encountered when studying atoms. The quantum number n, as we've used it, identifies molecular energy levels for the π electrons in molecules only.

Let us now consider what the ground state of vitamin A looks like. This molecule has ten carbon atoms and hence ten π electrons in the conjugated π bond system. These ten π electrons will try to occupy the lowest energy levels which are available to them. Because of the Pauli Exclusion Principle, which states that two electrons with the same spin cannot exist in the same state at the same time, each energy level can carry a maximum of two electrons having opposing spin. The ground state of vitamin A, therefore, has five filled

energy levels which are E_1, E_2, E_3, E_4, and E_5. Arranging these vertically with the largest energy at the top we would have the diagram of Figure 4-10.

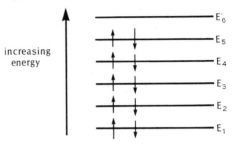

Figure 4.10

π electron energy electron levels in vitamin A

The lowest energy transition which ultraviolet or visible light could induce is a transition between E_5 and E_6. This doesn't mean that other transitions couldn't take place, it just means that they would require more energy.

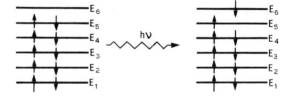

Figure 4.11

Lowest energy transition in vitamin A

From Equation (4-7)

$$E_6 = 1.19 \times 10^{-18} \text{ joules}$$

$$E_5 = 0.83 \times 10^{-18} \text{ joules}$$

The photon energy ($h\nu$) which could bring about this transition is $E_6 - E_5$. Thus,

$$E_6 - E_5 = \Delta E = h\nu = \frac{hc}{\lambda}$$

$$\lambda = \frac{hc}{\Delta E}$$

$$= \frac{6.63 \times 10^{-34} \times 3 \times 10^8}{3.6 \times 10^{-19}}$$

$$= 5.5 \times 10^{-7}\, m$$

$$= 550\, nm$$

Vitamin A should therefore absorb light near the middle of the visible spectrum. Experimentally vitamin A has an absorption maximum at 375 nm. Our calculation is therefore a fair bit off. The assumption that the molecule is stretched linearly as in Figure 4-4 is primarily at fault here. The conjugated bonds in vitamin A are in fact angled at 120 degrees to each other as in Figure 4-12 which shows a segment of the chain.

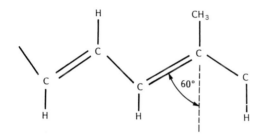

Figure 4.12

Segment of the vitamin A chain

A better estimate of ℓ the length of the molecule is $\ell = 9(0.15 \sin 60°) = 1.2$ nm. In addition theorists use fancier wave functions and are more careful about some of the other assumptions (i.e. assuming π electrons to be completely separate from σ electrons). Generally, however, their attack on the problem is similar to the one just discussed.

As a second and slightly more complicated example, consider light absorption by a porphyrin ring. These porphyrin rings are the chromophores (light absorbing groups) of chlorophyll, hemoglobin, myoglobin, and many other important biological molecules, playing vital roles in their functions. Figure

4-13 illustrates the porphyrin systems found in several different biological molecules (see page 50).

To reduce excessive computation these porphyrin rings can be considered simply as circles with radii of about 0.38 nm as shown in Figure 4-14. In this problem we have to fit electron wavelengths around the ring. The electron wave function $\psi(x)$ where x is a position on the ring must have the following property. If the electron starts off at some point x, having a wave function $\psi(x)$, then when it has gone once around the ring it has travelled a distance ℓ so that its wave function is now $\psi(x+\ell)$. But it has come back to its original starting point where its wave function is $\psi(x)$. Hence

$$\psi(x+\ell) = \psi(x)$$

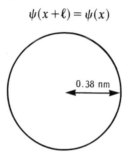

Figure 4.14

Circle model for the porphyrin ring

This property of periodicity is possessed by sine and cosine functions so that $\psi(x)$ can be either $\sin kx$ or $\cos kx$. Note that we cannot throw out $\cos kx$ as we did in the case of a linear chain of conjugated bonds because in this case it satisfies the boundary condition on the wave function. The condition $\psi(x+\ell) = \psi(x)$ must be satisfied so that

$$\sin k(x+\ell) = \sin kx$$
$$\cos k(x+\ell) = \cos kx$$

which means that $k\ell = 2\pi n$ where $n = 0, 1, 2, \ldots$. Thus it is necessary to fit whole wavelengths rather than half wavelengths on the ring (Figure 4-15) just as we did in finding

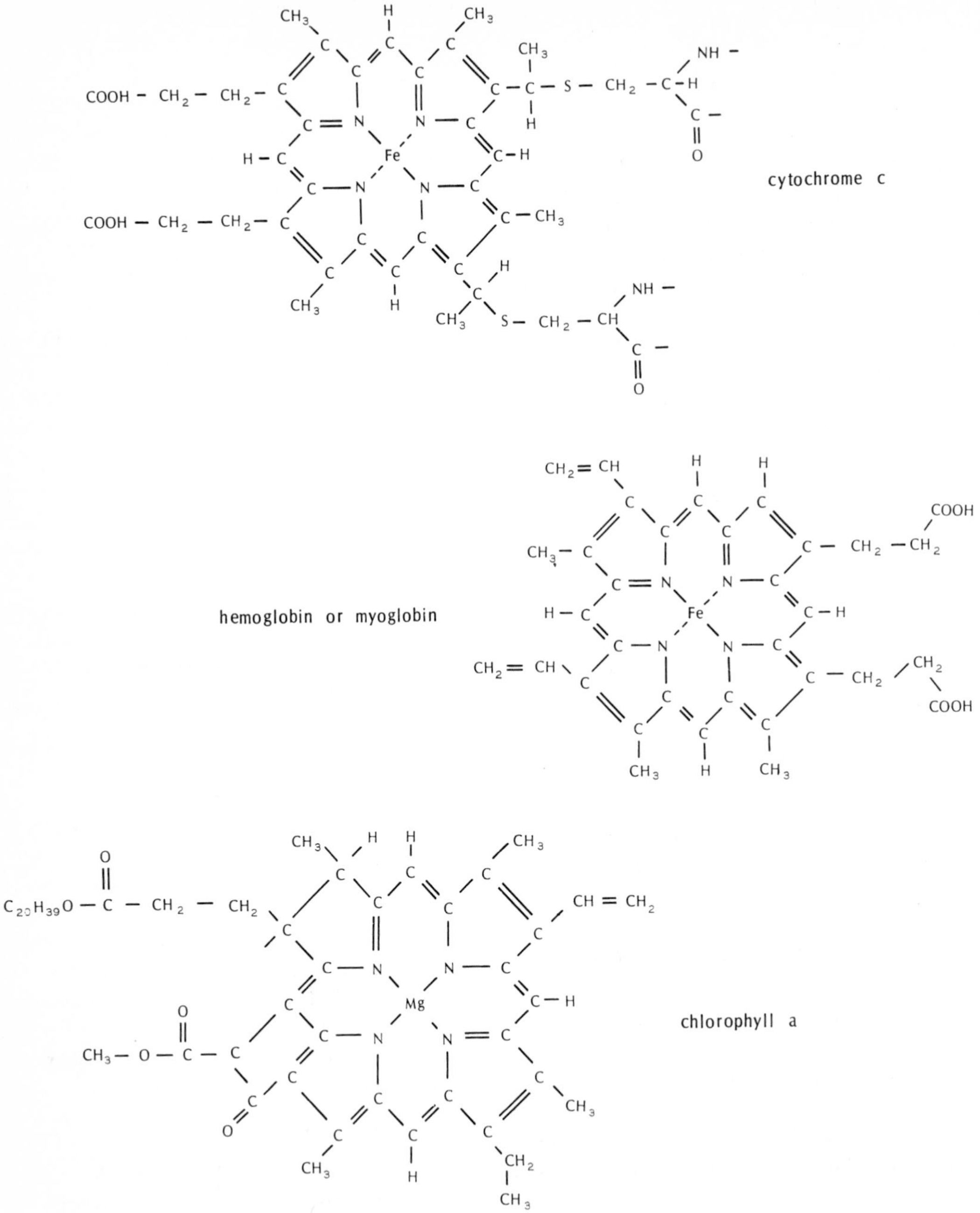

Figure 4.13
Several porphyrin rings from biomolecules

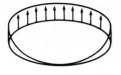
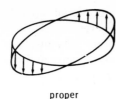

improper   proper

Figure 4.15
Improper and proper wave functions

the standing waves for a bell in Appendix 3. The properly normalized wave functions are

$$\psi_{sn} = \sqrt{\frac{1}{\ell}}\sin\frac{(2\pi n x)}{\ell}$$

$$\psi_{cn} = \sqrt{\frac{1}{\ell}}\cos\frac{(2\pi n x)}{\ell}$$

when $n = 0$, $\psi_{s0} = 0$. This means that the electron is never found in that state so that we can omit this solution. Thus there are two states when $n = 1, 2, 3, \cdots$ while for $n = 0$ there is one state,

$$\psi_{c0} = \sqrt{1/\ell}$$

The energy of each electron state can be determined as follows. The circumference of the ring corresponds to an integral number of electron wavelengths or

$$n\lambda_e = 2\pi r$$

$$= \ell$$

where r is the radius of the ring. From the de Broglie relationship.

$$\lambda_e = \frac{h}{m_e v}$$

thus

$$\frac{nh}{m_e v} = 2\pi r$$

or

$$\frac{n^2 h^2}{m_e^2 v^2} = 4\pi^2 r^2$$

Remembering that the kinetic energy, E_k, of the π electron is $\frac{1}{2}m_e v^2$ we can write, as before:

$$\frac{n^2 h^2}{m_e^2 v^2} = \frac{n^2 h^2}{2m_e\left(\dfrac{m_e v^2}{2}\right)}$$

$$= \frac{n^2 h^2}{2m_e E_n}$$

thus

$$\frac{n^2 h^2}{2m_e E_n} = 4\pi^2 r^2$$

$$E_n = \frac{n^2 h^2}{8\pi^2 m_e r^2} \qquad (4\text{-}8)$$

If we consider a specific example, chlorophyll, the empty (magnesium absent) porphyrin ring contains twenty π electrons. These twenty electrons fill up the states as shown in Figure 4-16. Note that the state with energy E_5 is only half filled, (the spin can point either up or down). Light can thus excite an electron from E_4 to E_5. The energy of the photon must equal the energy difference between these two levels.

$$\Delta E = E_5 - E_4$$

$$= \frac{h^2}{8m_e\pi^2 r^2}(5^2 - 4^2)$$

$$= 3.8 \times 10^{-19}\text{ joules}$$

The wavelength of the light is

$$\lambda = \frac{hc}{\Delta E}$$

$$= 523\text{ nm}$$

which is quite close to the actually observed value. When the magnesium atom is present as in Figure 4-13, two electrons from this atom occupy the vacancies in the fifth level. In this case, the lowest energy transition for the π electrons would be $n = 5$ to $n = 6$.

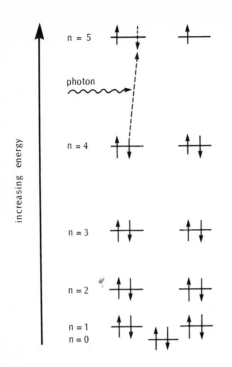

Figure 4.16

Lowest energy transition in the chlorophyll ring

In leaving the subject of UV-visible absorption by molecules it should be stressed that the only transitions studied so far are in long conjugated double bond sequences only. In molecules like the porphyrins the central ion, for example iron in haemoglobin and myoglobin, also absorbs visible light. The ligand field and crystal field theories used to describe the transitions in the central ion are similar to but considerably more difficult than the theory we have described here.

4-4 Absorption of Infrared Radiation by Molecules

Whereas UV-visible light causes transitions from one electron distribution to another in molecules, the absorption of infrared (IR) light leads to transitions between different vibrational energy levels. These vibrational energy levels arise because the molecular bond acts just like a combination of masses and springs and has, therefore, various natural frequencies of vibration, (Figure 4-17).

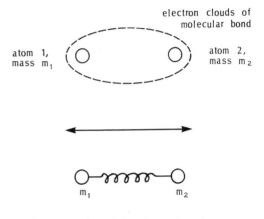

Figure 4.17

Molecular bond, mass spring analogy

The energies which correspond to the natural frequencies of vibration of such a system are given by the following equation

$$E_v = \frac{h}{2\pi} \sqrt{\frac{k}{\mu}} \left(v + \tfrac{1}{2}\right) \qquad (4\text{-}9)$$

The symbol v is an integer which indexes the vibrational energy levels. Its values are 0, 1, 2, 3, ... The symbol μ represents the reduced mass of the system defined by the equation

$$\mu = \frac{m_1 m_2}{m_1 + m_2}$$

where m_1 and m_2 are the masses of the atoms 1 and 2 respectively. The symbol k is called the force constant of the bond and is defined in the equation

$$F = -kx$$

where F is the force applied to separate the atoms and x is the distance they move (displacement). Transitions are allowed between adjacent levels only: that is $\Delta v = \pm 1$ always. From the form of the energy equation you can see that the energy levels are equally spaced.

Thus, transitions between any two adjacent levels can be induced by photons of the same energy. This contrasts strongly with the electronic energy levels considered earlier which are not equally spaced.

v		E_v
3	————————	$7/2\,(h/2\pi\sqrt{k/\mu}\,)$
2	————————	$5/2\,(h/2\pi\sqrt{k/\mu}\,)$
1	————————	$3/2\,(h/2\pi\sqrt{k/\mu}\,)$
0	————————	$1/2\,(h/2\pi\sqrt{k/\mu}\,)$

Figure 4.18
Vibrational energy levels

The fact that the vibrational levels are equally spaced plus the restriction that $\Delta v = \pm 1$ mean that each bond type will absorb IR light at one wavelength only. This absorption band has been an invaluable asset to chemists since molecules can be fingerprinted by their characteristic IR spectra. An abbreviated list of bonds and the corresponding IR wavelength absorbed is given in Table 4-2.

Table 4-2
IR Absorption Bands
for Some Chemical Bonds

Bond	Wavelength (μm)
C—H (aliphatic)	3.3–3.7
C—H (aromatic)	3.2–3.3
C—O	9.55–10.0
C═O (aldehydes)	5.75–5.80
S—H	3.85–3.90

We could go on and discuss other types of spectroscopy such as microwave spectroscopy involving transitions between molecular rotational levels. These energy levels are even more closely spaced than vibrational levels.

The basic principles involved are the same, however, namely that for a transition to take place, the energy of the absorbed photon must equal $E_{\text{final}} - E_{\text{initial}}$ where E_{final} and E_{initial} are the energies of the final and initial levels respectively.

4-5 Fluorescence and Phosphorescence

We are now in a position to consider the energy of a molecule as a whole; that is simultaneously consider electronic and vibrational energies. Before proceeding, however, it is necessary that we distinguish between electronic energy levels and electronic energy states. An electronic energy level specifies the energy of a particular electron. An electronic energy state, however, refers to the total electronic energy of the molecule; the sum of the energies of all the individual electrons present. Thus, in the electronic transition shown in Figure 4-19, the state on the left has all its electrons in the lowest levels possible and is called the ground state of the molecule.

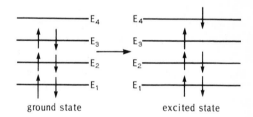

ground state   excited state

Figure 4.19
Typical electronic transition

The state on the right has only one electron which is in a level higher than the lowest one available to it. Nevertheless, the whole molecule is said to be in an excited state. From Figure 4-19 you should convince yourself that

$$E_{\text{excited state}} - E_{\text{ground state}} = E_4 - E_3$$

A molecule in the ground electronic state

will have many vibrational levels. Similarly, there will be many vibrational levels associated with the excited state. These states and some of their corresponding vibrational levels are shown in Figure 4-20.

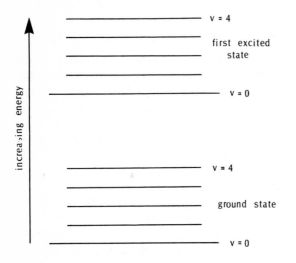

Figure 4.20
Energy diagram of ground and excited states

At room temperature most ground state molecules will be in the lowest vibrational energy levels. If a molecule does find itself in a high vibrational level it will, after several collisions, drop to a lower level and give up its excess energy to other molecules in its vicinity. We have, therefore, to distinguish between "hot" and "excited" molecules. "Hot" molecules are those which find themselves in higher vibrational energy levels than their neighbours. "Excited" molecules are those which are in an electronic state other than the ground state. A molecule can be in an "excited" state and yet be "cold" in the vibrational sense. These situations arise often when one considers the light absorption cycle shown in Figure 4-21. Initially, a photon of energy $h\nu$ is absorbed leading to the transition (1). The molecule is now both "hot" and "excited"

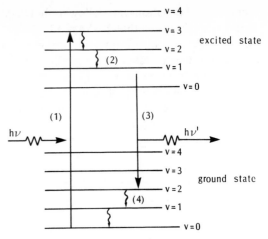

Figure 4.21
Origin of fluorescence

(this sounds like an episode from True Romance). It then collides with its neighbours, giving up its excess vibrational energy. The temperature of the system rises correspondingly. Once the molecule has dropped through its vibrational manifold to some lower level of the excited state, a process called vibrational relaxation (2), it can return to the ground state by radiating a photon of energy $h\nu$ (3). This process is called fluorescence. After more relaxation down the ground state vibrational manifold to the lower vibrational states (4), the molecule can be found at its original energy. The complete cycle takes about 10^{-7} seconds. There are several other important observations here,

(a) transitions can be from the ground state vibrational level to any of several vibrational levels of the excited state,
(b) the energy of the fluorescent photon must be equal to or less than the energy of the absorbed photon,
(c) some energy is lost in heating the system in the vibrational relaxation process.

Earlier when we considered the electronic energy levels of the excited state, we discussed energy level diagrams similar to that shown in Figure 4-22.

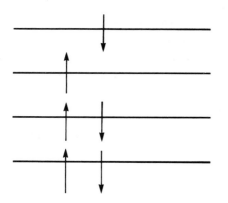

Figure 4.22
Typical excited state

So far, care has been taken to make sure that the spin of the excited electron is the same as it was in the ground state because the electron spin does not change during an electronic transition. Once the molecule is in the excited state, however, the spin of this electron can flip under the appropriate conditions.

The result of this spin flip is shown in Figure 4-23. The new state is called a triplet excited state. The singlet and triplet adjectives refer to the so-called multiplicity of the state. This multiplicity, M, is defined by the equation

$$M = 2S + 1 \qquad (4\text{-}10)$$

where S is the sum of the spin quantum numbers of the electron. Since the electron has spin quantum numbers of $+\frac{1}{2}$ or $-\frac{1}{2}$, it is simple to calculate the multiplicity. For the ground state,

$$S = +\tfrac{1}{2} - \tfrac{1}{2} + \tfrac{1}{2} - \tfrac{1}{2} + \tfrac{1}{2} - \tfrac{1}{2}$$
$$= 0$$

so that

$$M = 2(0) + 1$$
$$= 1$$

The ground state is thus a singlet state as is the first excited state. For the triplet state,

$$S = \tfrac{1}{2} - \tfrac{1}{2} + \tfrac{1}{2} - \tfrac{1}{2} + \tfrac{1}{2} + \tfrac{1}{2}$$
$$= 1$$

and

$$M = 2(1) + 1$$
$$= 3$$

The excited triplet state is slightly lower in

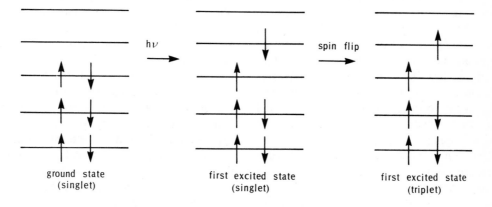

ground state   first excited state   first excited state
(singlet)   (singlet)   (triplet)

Figure 4.23
Singlet and triplet states

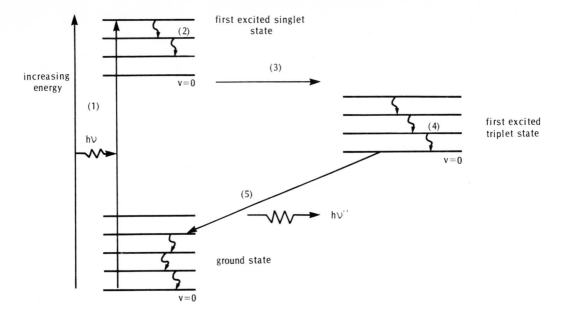

Figure 4.24
Origin of phosphorescence

energy than the corresponding singlet state. Let us now consider an example in which the triplet state is populated (Figure 4-24).

After the absorption of the incoming photon (1) the molecule begins to fall down, the vibrational manifold of the first excited singlet state (2). If, during this time, a spin flip occurs, the molecule suddenly finds itself in a triplet state (3). It can then fall down through the vibrational manifold of the triplet state until it reaches the $v = 0$ level (4). At this point the molecule is trapped. It cannot immediately fall to the ground state as happened in the case of fluorescence because it has the wrong multiplicity. (Transitions between states of differing multiplicity are forbidden.) It cannot return to the first excited singlet state because energy would be required. The molecule just sits there until some event occurring in its environment induces a spin flip. The molecule is then free to drop to the ground state and does so, simultaneously emitting light (5). This emission which can at times be several seconds after the original photon was absorbed is called a phosphorescence. Note that the phosphorescent photon, $h\nu''$, is of less energy than either the absorbed photon or the fluorescent photon.

While the excited singlet and triplet states are the sources of fluorescent and phosphorescent light respectively, they also are the intermediates in photobiological reactions. First of all, they have considerably more energy than the ground state and this excess energy is in just the range necessary to drive most chemical reactions. Secondly, the excited states have completely different physical properties than the ground state because the electronic charge has been redistributed. They can, therefore, take part in chemical reactions even though the ground state is often unreactive. Of the two excited state species, the triplet state is

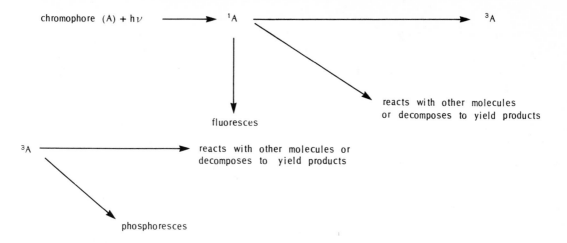

Figure 4.25
Possible processes in a photobiological reaction

involved in more photobiological reactions than is the singlet. This, in most cases, is simply a matter of time: the singlet state lifetime is only about 10^{-7} seconds and this state has little chance to react. The triplet state on the other hand is relatively long-lived and has plenty of opportunity to react. Figure 4-25 is a diagram of a flow chart of all the possible processes involved in a photobiological reaction.

The actual determination of which excited species, triplet or singlet, is involved in a specific photochemical reaction is sometimes clarified if one has a knowledge of the quantum yield ϕ, defined as follows:

$$\phi = \frac{\text{number of molecules altered}}{\text{number of photons absorbed}} \quad (4\text{-}11)$$

Values of ϕ considerably less than one are fairly common. A molecule may not react even though it has absorbed a photon; it could, for example, re-emit the photon instead. Usually, however, reactions which occur via the singlet state have a higher value of ϕ than reactions occurring via the triplet state. This is because one obtains a singlet state after every absorption whereas triplet states are produced less often. A great deal of other evidence is required before the knowledge of the photochemical pathway is complete. New techniques such as flash photolysis are beginning to provide more insight into the processes. The whole area of photochemistry and photobiology is developing into one of the most fascinating studies in science.

4-6 Biological Effects of Ultraviolet and Visible Radiation

The ultraviolet (UV) region of the electromagnetic spectrum (200–350 nm) is especially harmful to living systems. Cellular injury, mutations, and lethality have all been demonstrated to be the biological effects of intense doses of this light. In some cases, such as in the use of germicidal lamps, ultraviolet light can be used to sterilize rooms and equipment. However, great care must be taken in many clinical and industrial situations to insure that people are not exposed unwittingly to significant UV light levels. A quick reading of Chapter 8 may be useful if the student is unfamiliar with biological macromolecules.

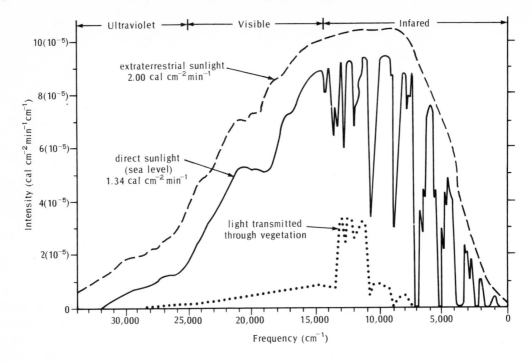

Figure 4.26

The intensity of the sunlight reaching the earth is not uniform but depends on the frequency, or colour, of the light. In addition, clouds and vegetation act as filters and transmit certain wavelengths preferentially. (Reproduced from D. M. Gates, "The Energy Environment in Which We Live," American Scientist, vol. 51, p. 327.)

Fortunately for systems living on earth very little of the intense solar UV light can penetrate the atmosphere. Figure 4-26 shows the spectrum of light in space and how it is modified by the atmosphere before it reaches the earth's surface.

The molecular aspects of UV induced damage in living systems have been an object of intensive study by photobiologists for many years. Central to their interest was the observation that deoxyribonucleic acid (DNA), the genetic carrier molecule, was involved. Some of the initial evidence for this was obtained by comparison of the "action spectrum" of bacterial cell death with the absorption spectrum (see Figure 4-27) of DNA. It was immediately

apparent that the wavelength of UV which was most efficient at bringing about cell death was precisely the wavelength which DNA absorbed most efficiently. Subsequently more detailed studies have shown that DNA can take part in several complex photoreactions. The results of these reactions, called photoproducts, are shown in Figure 4-28. Most celebrated and perhaps most damaging of all the photoproducts is the pyrimidine dimer. The formation of dimer, which corresponds to two individual pyrimidine molecules being joined by a new covalent bond, was first demonstrated in the late 1950s by Beukers and Berends who noticed their presence in ultraviolet irradiated frozen solutions of the

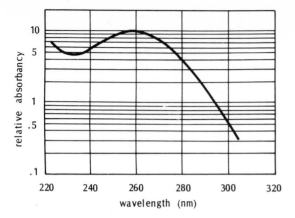

Figure 4.27
The absorption spectrum of DNA, showing the characteristic maximum at 260 nm. (from "Ultraviolet Radiation and Nucleic Acid" by R. A. Deering. Copyright© 1962 by Scientific American, Inc. All rights reserved.)

pyrimidine, thymine. Dimer formation in DNA was demonstrated soon after. Other experiments verified that dimer formation led to inactivation or lethality. The actual mechanism of inactivation by the dimer has been compared by P. C. Hanawalt, a noted photobiologist, to the effect one obtains if two adjacent teeth of a zipper become fused or clogged. The DNA is simply not able to function properly in such processes as transcription and replication.

The details of the photo-process leading to dimer formation are not completely understood. One scheme for uracil dimer formation has been proposed by J. G. Burr.

$$u \xrightarrow{h\nu} {}^1u^*$$

$$u + {}^1u^* \rightarrow u + {}^3u^*$$

$$u + {}^3u^* \rightarrow u - u \text{ (dimer)}$$

Both excited triplet and singlet states are present in this scheme.

A remarkable offshoot of the intense study of the formation of pyrimidine dimers in bac-terial DNA was the observation that under appropriate conditions damage of this type could be repaired by the cell. This new knowledge made a great deal of sense to many researchers who believed that significantly more UV light reached the earth's surface millions of years ago. The ability of the cell to combat some of the UV damage gave it a definite evolutionary advantage. One process of dimer repair is an enzymatic one involving recognition and excision of a length of DNA containing the damage. Alternate schemes for the process are shown in Figure 4-29.

Figure 4-28 indicates that several other photoproducts such as hydration products, chain breaks and protein crosslinks can be produced in DNA. Detailed information on these forms of damage and their relative importance are still matters of some conjecture, even though a great deal of effort is involved in their study. The interested student can follow the progress of the field by following such publications as *Advances in Photochemistry* and *Photochemistry and Photobiology*.

Unless the intensity is very high, such as one finds in a laser beam, visible light is essentially harmless to most living systems. There are some conditions and some diseases in which living systems can be "sensitized" to visible light. The result of this sensitization can, just as for ultraviolet light, be cellular injury, mutations, and lethality. One characteristic of these processes is that a sensitizer molecule, such as some dye molecules, absorbs the light to form an excited state. This excited state, which is usually a triplet state, then follows one of several possible mechanisms which lead to damage of a cellular molecule such as DNA or a protein. One often suggested mechanism is as follows

$$D \xrightarrow{h\nu} {}^1D$$

$${}^1D \rightarrow {}^3D$$

$${}^3D + R \rightarrow D + \text{damaged R}$$

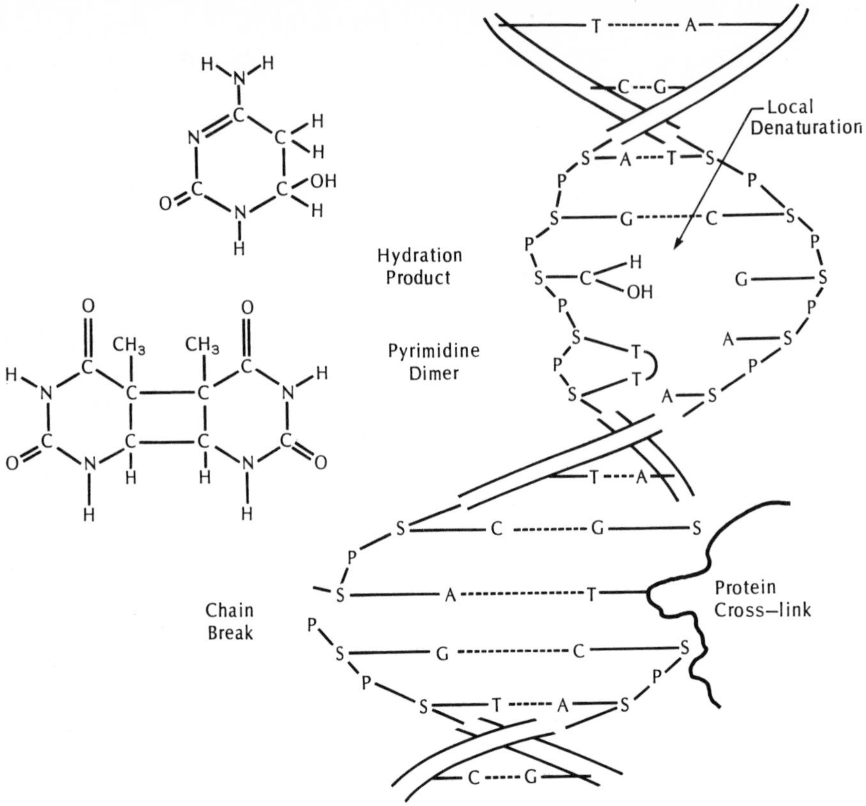

Figure 4.28

Schematic illustration of the various alterations found in DNA extracted from cells that have been irradiated with ultraviolet light. Some of these physical changes (e.g., denaturation) may be secondary effects resulting from photoproducts such as thymine dimers. The dimerized thymines are evidently unable to maintain hydrogen bonds with the adenine in the complementary strand. Other effects such as cross-linking to protein may involve new photoproducts yet to be characterized. Chemical structures for the cytosine hydrate and the thymine dimer are shown. (From "Ultraviolet Radiation and Nucleic Acid" by R. A. Deering. Copyright ©1962 by Scientific American, Inc. All rights reserved.)

In this scheme, D is the sensitizer molecule and R is the biological substrate. In one class of the reactions known as photodynamic reactions it is known that oxygen is involved. Damage in these cases can be especially severe and in humans the result is painful burns and sometimes death. In recent years certain tranquilizers, drugs, perfumes, and soaps have been found to contain powerful photosensitiz-ing agents. Many unfortunate experiences have been the result.

The last subject we should mention in this section is something everyone has experienced, sunburn. This condition occurs if one is exposed to a substantial amount of sunlight without the skin being pigmented (tanned). While the result is burns similar to those obtained from a photodynamic reaction, the

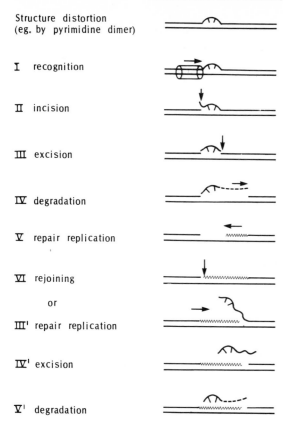

Structure distortion
(eg. by pyrimidine dimer)

I recognition

II incision

III excision

IV degradation

V repair replication

VI rejoining

or

III' repair replication

IV' excision

V' degradation

Figure 4.29
Schematic representation of the postulated steps in the excision – repair of damaged DNA. Steps I through VI illustrate the "cut and patch" sequence. An initial incision in the strand containing the damage is followed by local degradation before the resynthesis of the deleted region has begun. In the alternative "patch and cut" model, the resynthesis step III begins immediately after the incision step II, and the excision of the damaged region occurs when repair replication is complete. In either model the final step VI involves a rejoining of the repaired section to the contiguous DNA of the original parental strand. (reproduced from P. C. Hanawalt "Radiation Damage and Repair in Vivo," An Introduction to Photobiology, Carl P. Swanson, Ed. Prentice-Hall 1969.)

mechanism is entirely different. The action spectrum for sunburn is shown in Figure 4-30. The action spectrum indicates that the maximum erythemic (skin reddening) effects occur at approximately 295 nm, a wavelength which is present in the sun's rays, even at the earth's surface. The skin reddening which we notice is thought to be a result of vasoactive (affect vascular system) substances released in the epidermis by the radiation. Following the initial reddening, skin pigmentation or tanning occurs as melanin (the brown tanning substance) appears at the skin surface. The time course of erythema and pigmentation are shown in Figure 4-31.

4-7 The Poisson Distribution

Many processes in biological and physical systems depend on the random distribution of events. The absorption of light by molecules in a solution is an example which we will consider. First, however, we must become acquainted with a way of considering random events.

The Poisson distribution describes things that happen randomly and rarely. Imagine the following improbable situation. We have decided to observe buses passing a particular bus stop for a year and we know that during the hours when buses are running on the average, three pass in an hour. We happen to know the bus company, and they believe that the drivers should maintain a random schedule. There are many buses and many drivers and they start their routes whenever they feel so inclined. So these buses come by the stop randomly—at random times—and we have observed that on the average three go by per hour. How many will go by in a given hour? There is no answer to that question, since it might be any number; no bus may pass, or one bus might, or two, or any number might. But what we can do is tell what the chances are that a specified number of buses will pass the stop in the given hour and we can do this because the random arrival of the buses is governed by the

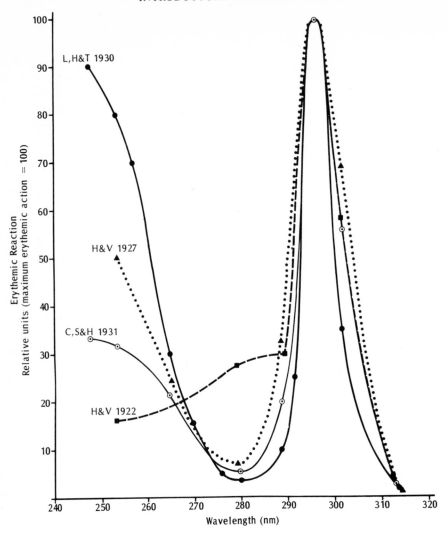

Figure 4.30

Action spectrum for erythema production. L, H, and T, 1930; Luckioesh, Holladay and Taylor (1930). H and V, 1927: Hausser (1928). C, S and H, 1931: Coblentz, Stair and Hogue (1931). H and V, 1922: Hausser and Vahle (1922). (from W. W. Coblentz, R. Stair, and J. M. Hogue, Bureau of Standards Journal of Research, 8, p. 541 (1932).) (reproduced from A. Wiskemann, "Effects of Ultraviolet Light on Skin," An Introduction to Photobiology. Carl P. Swanson, Ed.: Prentice-Hall Inc., 1969)

Poisson distribution. The probability that n buses will go by in a given hour, if the average number of buses passing in an hour is a, is

$$P(n) = e^{-a}a^n/n!, \qquad n = 0, 1, 2, \ldots \text{(4-12)}$$

$P(n)$ is the probability that there will be exactly n if the average is a. The probability that no bus passes is $P(0) = e^{-a}$. Here, $e = 2.718\cdots$. The symbol $n!$ is called n factorial and is the integer which equals the product of

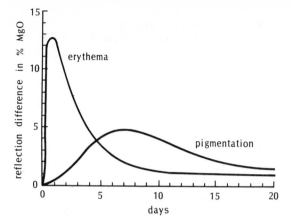

Figure 4.31
Temporal course of erythema and pigmentation after a 4 MED of ultraviolet light delivered by a hot quartz source. (adapted from Breit and Kligman) (reproduced from A. Wiskemann, "Effects of Ultraviolet Light on Skin," Introduction to Photobiology. Carl P. Swanson, Ed.: Prentice-Hall 1969)

n with the preceding integers:

$$n! = 1 \times 2 \times 3 \times \cdots \times n$$

For example,

$$4! = 1 \times 2 \times 3 \times 4 = 24$$

and

$$6! = 1 \times 2 \times 3 \times 4 \times 5 \times 6 = 720$$

Note that $0! = 1$, a case which arises when $n = 0$. In the example being considered, $a = 3$. What is the probability that exactly 3 buses will go by in a given hour?

$$P(3) = e^{-3}(3)^3/3! = e^{-3} \times 27/(3 \times 2 \times 1) = 0.225$$

or $22\frac{1}{2}$ percent of the time exactly 3 buses will pass in a given hour. The probability that no bus will pass is not zero!

$$P(0) = e^{-3} = 0.05$$

So 5 percent of the time not a single bus will pass, and if we calculate it, about 9 percent of the time, 6 buses pass. The following kind of question can also be answered:

What is the probability that two or fewer buses come by in an hour?

$$P = P(2) + P(1) + P(0)$$
$$= 0.225 + 0.150 + 0.050 = 0.425$$

or $42\frac{1}{2}$ percent of the time less than three buses go by. Suppose we need to know what the chances are that three or more buses go by. Does this mean that we must calculate $P(3)$ and $P(4)$ and $P(5)$ and so on ad infinitem and add them up to get the answer? You can try this, but there is a better way. The probability that any number (including zero) buses goes by is exactly one. That is,

$$1 = P(0) + P(1) + P(2) + P(3) + P(4) + \cdots$$
$$= \{P(0) + P(1) + P(2)\} + \{P(3) + P(4) + \cdots\}$$

That is, on moving the first three terms on the right over to the left,

$$1 - \{P(0) + P(1) + P(2)\} = P(3) + P(4) + \cdots$$

Thus, the probability of three or more going by is just 1 minus the probability that 2 or fewer go by, and from the calculation above this is

Probability that 3 or more go by $= 1 - 0.425$
$$= 0.575$$

or 57.5 percent of the time.

It should not be difficult for you to imagine that instead of buses going by a stop randomly, we might be talking about the number of photons in a flash of light, or the number of photons coming from a faint star in a given second, or the number of radioactive decays which might actually take place in a given hour, or the number of goals Team Canada might make in their next hockey game with the Russians, assuming they make their goals randomly

4-8 The Beer-Lambert Law: Spectrophotometry

The amount of light absorbed by a solution, a solid or even tree leaves can be expressed in two ways. The first is called transmittance, usually given the symbol % T.

$$\% \, T = \frac{I}{I_0} \times 100 \qquad (4\text{-}13)$$

where I_0 is the intensity of the beam before it passes through the sample, and I is the intensity after passing through the substance or the sample. A second, and more important scale, is called absorbance ($\mathscr{A}$) or sometimes optical density (O.D.).

$$\mathscr{A} = \text{O.D.}$$

$$= \log \frac{I_0}{I} \qquad (4\text{-}14)$$

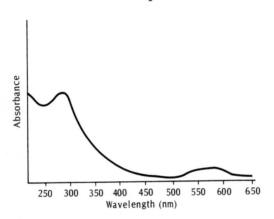

Figure 4.32
Absorbance spectrum of metmyoglobin

The popularity of this scale stems from the Beer-Lambert law which states that the absorbance of a sample depends upon its overall thickness (ℓ in centimetres) and the concentration (c in mol l^{-1}) of the light absorbing species, the chromophore. In mathematical form this is

$$\mathscr{A} = \log \frac{I_0}{I}$$

$$= \varepsilon c \ell \qquad (4\text{-}15)$$

where ε is a proportionality constant, which is termed the extinction coefficient, and has the units $l\,\text{mol}^{-1}\,\text{cm}^{-1}$. The value of ε differs for every wavelength and every chromophore. Usually, however, extinction coefficients of chromophores are quoted at the wavelengths which are most strongly absorbed. These correspond to the peaks of an absorption spectrum such as shown in Figure 4-32.

The origin of the Beer-Lambert law is related to the random distribution of chromophores in the solution. Such a solution is shown in Figure 4-33, with the size of the light absorbing species grossly exaggerated. If

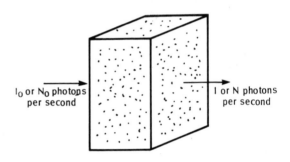

Figure 4.33
Solution containing a random distribution of chromophores

a photon passes through one of the many spaces of the solution it emerges unchanged in any way. If, however, it passes very close to a chromophore the chances are very good that it will be absorbed. Whether or not the photon will be absorbed depends on the area presented by the chromophore (a) and its ability to capture the photon (p). The product of these two terms is called the effective cross-section s. Thus

$$s = ap$$

Note that s is thus a characteristic of the chromophore and is, as we'll see, directly related to ε, the extinction coefficient. The average number of photons absorbed by the sample depends on s, the concentration of the chromophore (c) and the thickness of the solution (ℓ).

Average number of photons absorbed $= N_{av}$

$$= sc\ell$$

The probability that the photons will get through the solution is the probability that .they will not be absorbed. This corresponds to the $P(0)$ case of the Poisson distribution mentioned in the previous section.

$$P(0) = e^{-N_{av}} = e^{-sc\ell}$$

but

$$P(0) = \frac{n}{n_0} = \frac{I}{I_0}$$

thus

$$\frac{I}{I_0} = e^{-sc\ell}$$

or

$$ln\frac{I}{I_0} = -sc\ell$$

This can be rearranged to yield

$$ln\frac{I_0}{I} = sc\ell$$

Changing this to a logarithm to the base 10 results in

$$\log\frac{I_0}{I} = 0.4343 \, ln\frac{I_0}{I} = 0.4343\,sc\ell$$

The extinction coefficient ε is defined to be $0.4343s$. Thus

$$\log\frac{I_0}{I} = \varepsilon c\ell$$

and the Beer-Lambert law is verified.

If two chromophores or more exist simultaneously in the same solution and they do not react, then their absorbances may be added. Thus

$$\mathscr{A}_{solution} = \mathscr{A}_1 + \mathscr{A}_2 + \cdots$$

$$= (\varepsilon_1 c_1 + \varepsilon_2 c_2 + \cdots) \quad (4\text{-}16)$$

This equation is often useful in dealing with biological solutions, which often have many components.

light source	mono chromator	sample	detector	meter

Figure 4.34

Block diagram of a spectrophotometer

The device which measures the amount of light absorbed by a sample is called a spectrophotometer. These instruments come in many shapes and sizes and the numbers printed on their price tags are just as variable. Basically, however, a spectrophotometer is composed of the five parts shown in Figure 4-34. The light source which is used depends on the wavelength region of interest. Hydrogen lamps are usually used for the ultraviolet spectrum and tungsten lamps are used for the visible spectrum. The monochrometer which may be either a prism or a grating serves to divide the light from the source into its component wavelengths.

A prism is familiar to all of us, but a grating may be a new word for some. Essentially it is a plate containing very closely spaced lines. When a light beam is transmitted through or reflected from a grating one obtains a dispersion or separation of the component wavelengths as in Figure 4-35. The detector is a light-sensitive device such as a photocell, a photodiode or a photomultiplier. Typically such a device works by producing an electrical current which is proportional to the light intensity falling on it. The meter usually is designed so that either absorbance or percent transmittance can be read. In the more sophisticated devices, the wavelength setting of the monochromator can be increased or decreased automatically and a chart recorder is used instead of a meter. The most expensive and

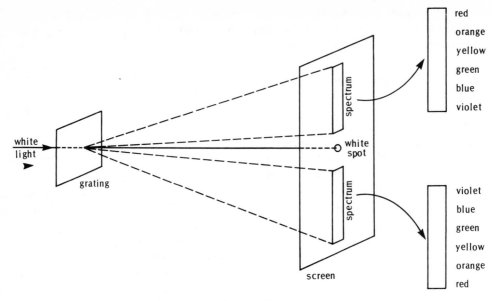

Figure 4.35
Dispersion by a grating

elaborate spectrophotometers are often on-line to a computer which can be programmed to control the whole operation and monitor absorbance or percent transmittance automatically.

The study of light absorption and emission is a vast subject and only a small introduction has been given in this chapter. With the invention of the laser, many new and useful biophysical techniques have emerged. For further reading the student is referred to the bibliography at the end of the text.

Problems

1. A biophysicist is using a beam of neutrons to study structural changes in connective tissue brought about by muscular dystrophy. In these experiments it is very important that the wavelength of the neutrons is known. By a technique called time-of-flight, he determines the velocity of the neutrons to be $1.3 \times 10^3 \, \text{m s}^{-1}$. What is the wavelength of the neutrons? (The mass of the neutron is $1.67 \times 10^{-27} \, \text{kg}$.)

2. An electron at a certain energy level in a molecule can jump to an adjacent level 6×10^{-19} joules higher. Will the molecule containing this electron absorb light in the visible region (400–800 nm)?

3. You are given two distinct electronic energy levels to be occupied by two (indistinguishable) electrons. Sketch all those diagrams which show ways of doing this that are consistent with the Pauli Exclusion Principle.

4. Ultraviolet light in the 260 nm region is especially damaging to the DNA of cells. How much energy is involved when a DNA molecule absorbs a photon of this light?

5. Consider a π electron in the $n = 2$ molecular energy level of butadiene. The wave function for this situation is

$$\psi_2 = \sqrt{\frac{2}{0.45}} \sin \frac{2\pi x}{0.45}$$

Butadiene may be sketched as

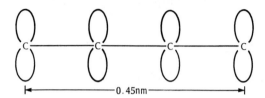

and the wave function looks like

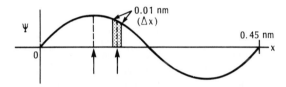

What is the probability density at the points indicated by the arrows? What is the probability of finding an $n = 2$ electron in the shaded region? Sketch the probability density in the $n = 2$ and $n = 3$ energy levels.

6. What is the energy of the highest filled electronic molecular orbital in the ground state of butadiene?

7. Below is the molecule 1,3-hexadiene, and those carbons which comprise a conjugated system are numbered

$$CH_2 = CH - CH = CH - CH_2 - CH_3$$
$$1 \quad\quad 2 \quad\quad 3 \quad\quad 4$$

C—C bond length = 0.15 nm in conjugated system

What is the expression for the wave function of the second electronic level? For an electron in this level, what is the probability density at the second carbon atom?

8. Consider an electron in the $n = 1$ electronic energy level of retinal. What is the probability density of the electron at carbon four? The C—C bond length is 0.15 nm. Sketch the wave function in the $n = 2$ energy level.

9. In a certain linear, conjugated π system the $n = 5 \rightarrow n = 6$ transition is induced by light of wavelength 548 nm. What is the length of the system? How many C—C bonds does it contain? (Each is of length 0.15 nm.)

10. Absorption maxima are often associated with the absorption of a photon by a π electron, after which the π electron is left in an excited state. How do you explain the fact that the absorption spectrum covers a whole range of wavelengths, including λ_{max}, rather than being sharply confined to a very definite wavelength, corresponding to the energy difference between the π electronic states involved?

11. A molecule in which there is a sequence of conjugated double bonds 12 atoms long is observed to have a maximum absorption at 500 nm. For a similar molecule, in which the conjugated sequence is only 8 atoms long, at what wavelength would you expect to find the maximum absorption?

12. An atomic absorption spectrum consists of individual lines. Why then for a complex molecule does the absorption spectrum consist of broad bands?

13. Let E_r represent the minimum energy required to bring about a transition between adjacent rotational energy levels in a molecule. E_v will represent the

minimum energy required to bring about a change between vibrational energy levels in a molecule and E_e will represent the minimum energy to bring about a change in electronic configuration. What is the correct ordering of these energies from the largest to the smallest?

14. Molecules may become excited by absorbing electromagnetic radiation. The absorbed energy may be used to (i) increase the rotational energy of the molecule, (ii) increase the vibrational energy, (iii) place the electrons in a different distribution. If a molecule absorbs a photon of ultraviolet radiation which of (i), (ii), or (iii) will occur?

15. In order to obtain a standing wave on a ring, how many half wavelengths must there be around the circumference?

16. Calculate the wavelength of light which would induce the lowest energy transition of benzene. The C—C bond length is 0.15 nm.

17. A solution of biological molecules absorbs 380 nm light most strongly and fluoresces most strongly at 610 nm. What is the amount of energy per photon lost in the process?

18. The average number of mice caught by a bobcat is 4.3 per day. What is the probability that on a particular day the bobcat will catch more than 2?

19. Pierre is portaging with his canoe and finds he gets 300 black fly bites in a 1 hour 40 minute portage. For how many one minute intervals was he free of black fly bites?

20. Trapped in an airport during a thunderstorm a statistician decides to count lightning flashes. He notes the flashes in each 1 minute interval during the storm and gets the following data:

# of Flashes	Frequency
0	2
1	7
2	7
3	4
4	2
5	1

(a) Calculate the average number of flashes per interval.

(b) Assuming the data fits a Poisson distribution calculate the expected frequency of 0, 1, 2, 3, 4, 5 lightning flashes per 1 minute interval in a similar 23 minutes.

(c) Is the assumption of a Poisson distribution a good one?

21. The absorbance of a solution is 0.6. What is the percent transmittance?

22. A solution of a particular chromophore absorbs strongly at wavelengths around 530 nm (green). If this solution is looked at in white light what color will it not appear to be?

23. Suppose that the percent transmission of a given solution of length 1 cm is 50 percent. If the length of the solution in the light path were three times larger what would the percent transmittance be?

24. About 95 percent of the photons striking a sample of tree leaves is absorbed. What is the absorbance and the percent transmittance of the leaves?

Chapter 5
Quantum Nature of Vision

5-1 Introduction

In Chapter 3, the optical properties of the eye were investigated to show how an image is formed on the retina. In this chapter we will investigate how that image is recorded, that is we will investigate the actual mechanisms of vision itself.

5-2 Physiology of the Retina

A schematic drawing of the retina is shown in Figure 5-1. The pigment epithelium, or pigmented layer, contains a dark pigment which absorbs light, much as the black paint within a camera absorbs stray light which would otherwise destroy the optical image. But notice that the receptors (the cells next to the pigment epithelium) are as far away from the source of light as possible; in other words, the retina is built wrong side forward. In vertebrates this occurs because the eye is an outgrowth of the brain. The light-sensitive part of the receptors points away from the source of light. Light must traverse the intermediate nerve cells and fibres and even the cell bodies of the receptors before reaching the light sensitive parts. Although the nervous material is almost transparent, it is not uniformly so. It bends light and the optical image, already fuzzy because of imperfections in the other material of the eye, is even further degraded. As so often happens in biological systems, we have here a poorly designed device which functions brilliantly. You may be surprised to learn that the octopus, an invertebrate, has its eye built the "right" way round.

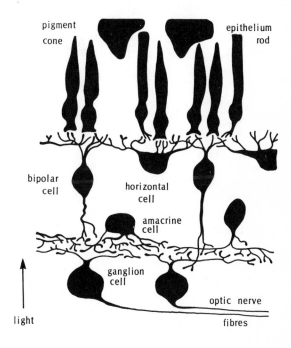

Figure 5.1
Schematic drawing of the retina (adapted from F. S. Werblin "The Control of Sensitivity in the Retina" Scientific American **228** (1), 1973.)

Figure 5-2 shows a light micrograph of a cross-section of the retina in the region of the fovea. When we want to look directly at an object we move our eyes until the image of that object lies on the fovea. In ordinary light, this is the region where vision is best. For the micrograph the retina was stained so that some of the cell bodies appear as black dots; in the living eye, as has been mentioned, this

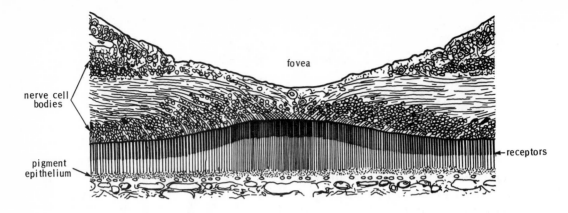

Figure 5.2

Cross section of the retina (from S. Polyak, "The Vertebrate Visual System," University of Chicago Press)

tissue is transparent. The fovea is about 0.1 mm thick, while the rest of the retina is about twice that thickness. The receptors located in the fovea are called cones. They are responsible for our colour vision, and for our very accurate daytime vision. There are approximately 6 million cone cells in the human eye, the vast majority of them located in the fovea whose diameter is about 0.3 mm. The other type of receptor is the rod cell. Almost all of these are located outside the fovea. There are approximately 120 million of these. The rods are sensitive to lower intensities of light than are the cones (which are inoperative at night). It is the rods, therefore, which are responsible for our night vision.

Figure 5-3 is a drawing made from electron micrographs of rod cells. It is in the outer segment of the rod cell that the light sensitive material is contained. The outer segment is a long cylindrically shaped membrane filled with "disks," which are stacked on top of one another. The outer segment has a diameter of about 1 or 2 μm (compare that with the wavelength of visible light, which is about $\frac{1}{2}$ μm), while its length is about 10 times that.

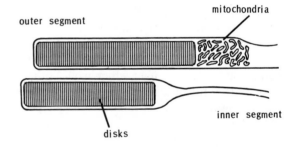

Figure 5.3
Rod cell

The disks are also composed of membrane. Figure 5-4 shows a cross-section of a rod cell, showing the disk membrane which is continuous and encloses an inside space. The disk is similar to a completely deflated basketball. The membrane making up both sides of the disk is believed to have a molecular structure typical of many membranes. It is composed of two molecular layers, each layer consisting of molecules called phospholipids. These molecules consist of a "head" group containing a phosphate and it is hydrophilic. Attached to the hydrophilic head are two fatty

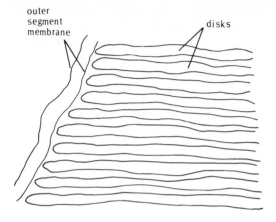

Figure 5.4

Cross section of a rod cell

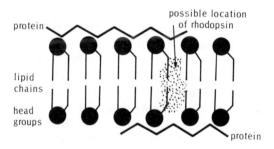

Figure 5.5

Bilayer membrane

chains, the lipids. These chains are hydrophobic and thus the membrane may look like the one pictured in Figure 5-5. See Chapter 8 for a discussion of hydrophilic and hydrophobic interactions.

There is some protein associated with the upper and lower surface. Situated somewhere within this membrane, between the phospholipids are the large molecules of the visual pigment. In the case of rod cells, this pigment is called rhodopsin. The diameter of the rhodopsin molecule is about $4 \times 10^{-3} \mu m$.

5-3 Mechanism of Vision

Rhodopsin is a combination of retinal and a large protein called opsin. Retinal (Figure 5-6) is very similar in structure to vitamin A. A line protruding from the structure in Figure 5-6 indicates the position of a methyl group. Retinal differs from vitamin A in having two fewer hydrogens (it is the aldehyde corresponding to vitamin A, which is an alcohol, also called retinol). A long molecule such as this can exist in several different forms. The form shown in Figure 5-6, in which the single and double bonds jog up and down, keeping the carbon tail pointed directly away from the ring, is called all-trans retinal. It is not the form which is combined with opsin to form rhodopsin. That isomer (meaning that it has the same chemical formula but a different physical shape) is shown in Figure 5-7.

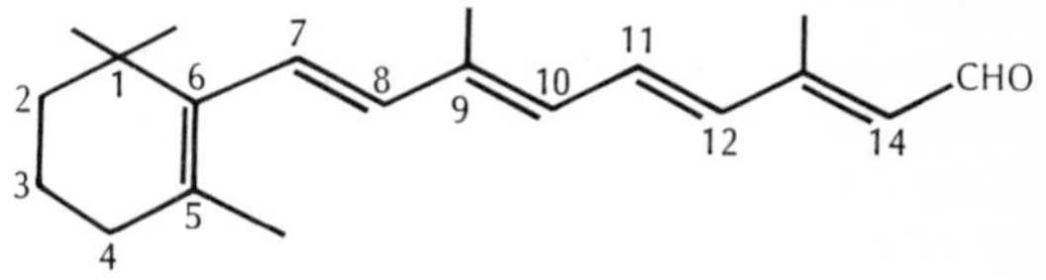

Figure 5.6

Trans-retinal

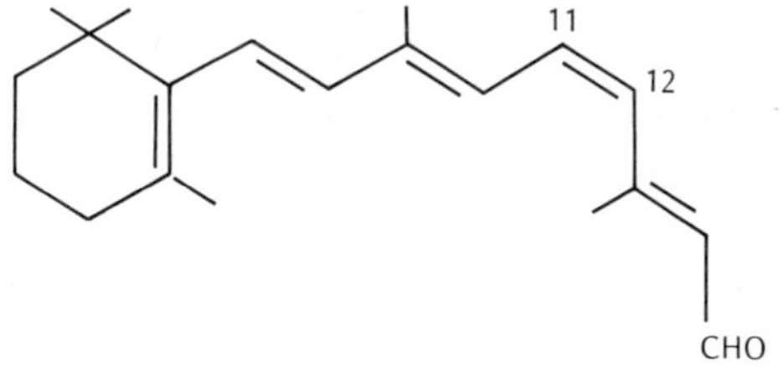

Figure 5.7

Cis-retinal

This form is called 11-cis retinal. It is the same as the all-trans form except for having been twisted about the double bond located between the 11th and 12th carbons. When

11-cis retinal absorbs light, it straightens out to become all-trans retinal. It does this after a π electron makes a transition to an excited level. This allows the molecule to straighten out and rotate about the line joining the 11th and 12th carbons. After it is in the all-trans configuration, retinal detaches itself from opsin. It hydrolyses to become vitamin A and migrates to the pigment epithelium, where it undergoes enzymatic regeneration to the 11-cis form. From the pigment epithelium, it can migrate back to the outer segments and combine with opsin in one of the disk membranes. When retinal is in the 11-cis form it binds readily with opsin. These bonds are broken after the absorption of light. Exactly what bonds are involved is still a matter of research. There is good evidence that the principal one is a C—N bond involving the 15th carbon of retinal and one of the nitrogens of a lysine group of opsin.

The time course of regeneration of rhodopsin in the eye has been investigated. Suppose at zero time a light strong enough to isomerize all of the pigment molecules is extinguished. The proportion of pigment molecules in the isomerized state is found to decay exponentially with a half-life of 5 min (i.e. half of the molecules have regenerated during the first 5 min after the light is extinguished). This regeneration process is thus characterized by the following familiar expression

$$N = N_o e^{-kt} \tag{5-1}$$

where N/N_o is the proportion of molecules in the isomerized state and k is a constant defining the particular process. For rhodopsin $k = 0.14$ min^{-1}.

Immediately after the adapting light is turned off, the pigment molecules are all in the separated state. Thus further light shone on the system will elicit no response and we say that the threshold intensity in this situation is infinite. After 5 min, half of the

molecules have regenerated but it turns out that the threshold intensity does not vary inversely with the number of pigment molecules regenerated. Rather, the threshold intensity obeys the following relationship

$$\log \frac{I}{I_o} = 20C \tag{5-2}$$

where I is the threshold intensity, I_o is the threshold intensity after the eye has been dark-adapted, C is the fraction of pigment molecules in the normal state. Note the similarity between Equation (5-2) and the response-stimulus function for sound intensity that was developed in Chapter 2.

We have looked at the structure of the receptors and have seen approximately where the visual pigment is located. Even though the functioning of the receptor is still pretty much a mystery, certain characteristics of the visual pigment can be determined rather accurately. The molecule rhodopsin can be extracted from rods and we can measure its absorption spectrum. We would expect rhodopsin to absorb light most at some particular wavelength. First of all, not all electromagnetic radiation is visible light. Visible light has wavelengths between 720 and 400 nm only. That means that rhodopsin should absorb within this range of wavelengths and the absorption should be small at the long and short wavelengths. This is almost true, and in Figure 5-8 we see the absorption (or action) spectrum of rhodopsin. It has a peak at about 510 nm (a yellowish green). Thus, if rhodopsin is viewed in white light, it will appear reddish. The reflected light contains those wavelengths which have not been strongly absorbed; it contains red and blue, the yellow and green having been absorbed by the rhodopsin. After rhodopsin has been exposed to light so that a large fraction has absorbed light and has broken down as a consequence into opsin and retinal, the reflected light appears yellowish white. Thus, we say that light bleaches the visual pigment.

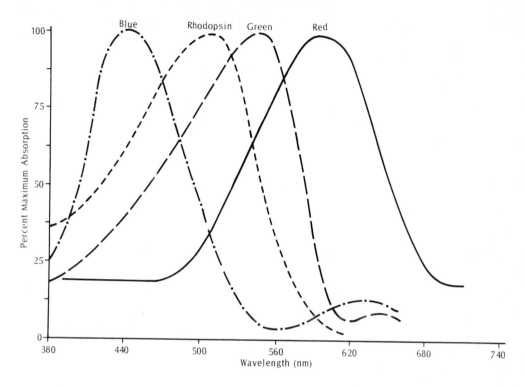

Figure 5.8
Action spectra of rhodopsin and the colour pigments in the cones

5-4 Colour Vision

The cone cells contain different pigments and these are responsible for colour vision. Perhaps you are familiar with the primary colour theory which attempts to describe all colours as a mixture of three primary colours. If we are concerned with the colours which arise on mixing pigments, the primary colours are red, yellow, and blue. Thomas Young, around the beginning of the last century, claimed that colour vision in humans could also be understood if there were three visual pigments with colours red, blue, and green. Is there any correspondence of this theory with what we now know about colour vision? First, it should be mentioned that, as opposed to rhodopsin, the pigments of the cones are not

easily extracted. To obtain their absorption spectra, we must proceed using white light and analysing the reflected light, as discussed above in the case of rhodopsin. Since there are quite possibly several pigments, the resulting action spectrum for all the cones together will be more complicated than for just one kind of receptor cell. Thus it is necessary for the experimenter to extract from his data the action spectrum for each kind of pigment. When the wavelengths of the reflected light are analysed and the total action spectrum of all the cones is deduced and then resolved into the action spectra of the individual pigments, we find that there are three pigments, and thus possibly three distinct kinds of cone cells, with absorption maxima at about 440, 530, and 570 nm. Figure 5-8 is a typical set of action spectra deduced in this way. All of the peaks have been arbitrarily set at 100 percent when compared with one another. Also included for comparison is the action spectrum of rhodopsin which has been normalized in the same way.

5-5   Visual Sensitivity

As the light intensity is reduced, as happens naturally at twilight, cone (day) vision gives way to rod (night) vision. On a clear night, in rural areas, stars which are extremely faint are visible to the human eye. Astronomers for thousands of years have known that the eye can detect very faint amounts of light and only comparatively recently have we had instruments which could discriminate between magnitudes of faint stars as well as our eyes could. Nowadays we might phrase a question related to this in the following way: How few photons of visible light does the eye respond to? Here we should distinguish two questions. First, how many rods must respond before a visual event is sensed; and, secondly, how few photons are necessary for a rod cell to respond. The answer to the latter question is that the absorption of only one photon will elicit a response from the rod. Fewer than 10 rods responding within a short time of one another will cause us to experience a visual event. In the 1940s a series of experiments was carried out by Hecht, Schlaer, and Pirenne which gave these fundamental and somewhat surprising results. We will make extensive use of the Poisson distribution of random events which was discussed in Section 4-7.

Hecht, Schlaer, and Pirenne carried out experiments on human subjects whose eyes had been dark-adapted before the experiments began. It takes about 40 minutes in a darkened room before the eye is dark-adapted and at its most sensitive level. In the dark, the human eye is most sensitive to the wavelengths around 510 nm, which is rhodopsin's absorption maximum. There were two other things the experimenters had to know before they could proceed to determine the ultimate sensitivity of the eye. How long could a flash of light be on before the eye would get used to it, and thus spoil the sensitivity determination? How large a spot could be illuminated on the retina, before the same sort of thing happened? They found that the flash had to be $\frac{1}{10}$ or less seconds in length, and fall on an area that contained only about 500 rod cells. Their apparatus was designed so that it delivered a flash of duration $\frac{1}{10}$ s at a wavelength of 510 nm to an area of the retina containing about 500 rod cells. By decreasing the intensity of the light they could determine when the eye would fail to see the flash and thus determine the threshold of seeing. As a result of their experiments they found that about 100 photons on the average in a flash was just at the threshold. How bright would such a light source be? Another way to ask the question is to ask what wattage bulb would produce photons at the rate of 100 per $\frac{1}{10}$ s, or, in other words, at the rate of 1000 per second? Since the wavelength is 510 nm, we can multiply the number of photons per second by the energy of one photon:

$$\text{photons s}^{-1} \times \text{joules per photon} = \text{J s}^{-1}$$
$$= 1000 \times hc/\lambda$$
$$= 4 \times 10^{-16} \text{ watt}$$

That sort of bulb is not available at the corner store! In fact, what must be done is to use an ordinary light source in conjunction with filters to cut down the amount of light.

The right kind of filter to use is one which treats all wavelengths alike. Such a filter is called a "neutral" filter. It is characterized by its absorbance (or density, as it is called in the filter trade).

$$\mathcal{A} = \log I_o/I$$

For example, if the density of a filter is 1.0, it filters out all but 10 percent of the incident light. Therefore

$$I = I_0/10$$

If two of these filters are put together, the first allows only $\frac{1}{10}$ of the light through and the second allows only $\frac{1}{10}$ of that light through, so, two density 1 filters together pass only $(\frac{1}{10} \times \frac{1}{10})$ or $\frac{1}{100}$ of the original light. That is, two density 1 filters together make a density 2 filter. By putting together various filters, the original intensity can be reduced to almost any amount. This is what they had to do. Since they knew the total density of the filters they used and the intensity of the light coming to the filters from their source, they could calculate the intensity of light leaving the filter system, which was then directed onto the retina. Thus they knew the average number of photons per flash coming out of their system corresponding to the threshold of vision, and this number was about 100.

Not all of those 100 photons get to the receptors and activate them, however. There is quite a decimation of photons in the eye. A small fraction are reflected from the cornea and the lens surface. About $\frac{1}{2}$ are absorbed or deflected in the vitreous humour. By the time they arrive at the retina, the number is down to about 45. Then they enter the retina. There, 80–90 percent are absorbed in the nerve fibres, nerve cells, and other material. Ten or fewer are left which can then activate the rod cells. The exact number is not known—these experimenters said 10 were necessary for the flash to be seen, while more recent experimenters have said that as few as two photons on the average activating receptors cause a positive response from the subject. Now, whether the number is 2 or 10, one thing is certain. If even as many as 10 fall on the 500 rod cells, it is pretty sure that most of the time each rod which is activated is activated by only 1 photon.

Suppose that the threshold of vision corresponds to flashes in which, on the average, 5 photons activate rod cells. We can be certain that, after all that filtering, the photons which get through are random and so obey the Poisson distribution (see Chapter 4). Similarly the number of rods activated per flash (which, we are assuming, is 5 on the average) is governed by the Poisson distribution. So we can compute the percentage of flashes in which say 3 rods are activated, if 5 are activated on the average. Figure 5-9 shows graphs of the Poisson distributions for averages of 1, 5, and 10.

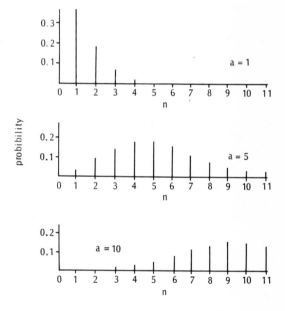

Figure 5.9
Poisson distributions for activation of the rod cells

At each integer, a line has been drawn so that its length is the value of the probability of that exact number of events occurring. Remember that

$$P(n) = e^{-a}a^n/n!$$

You can see how their shape changes with average number. For example, if the average number is very low e.g. 1, most flashes will contain a very small number of photons. For $a = 1$, $P(0) = e^{-1} = 0.368 = P(1)$. When $a = 5$, $P(0)$ is less than 0.01, while for $a = 10$ it is almost zero. In this graph, if we add the lengths of these lines, the total should be 1. That is, the total probability of something or nothing happening is 1.

Now suppose that if 5 or more rods are activated in one of these flashes, we will report that we have seen a flash, while if fewer than 5 are activated we will say that we did not see it. Everything to the right of 5 and including 5 corresponds to a yes response and everything to the left of five corresponds to a no reply. We see that almost none of the flashes will give a yes response if a is 1, while almost all of them will if a is 10. If a is 5, some fraction between zero and 1 of the flashes will give a positive response. A better way to present this is to plot the fraction of flashes seen as we vary the average number of photons per flash. We are still assuming that the number of rods needed to see the flash is at least 5. What is plotted in Figure 5-10 is the probability that 5 or more rods are activated per flash, for various average numbers of rods activated per flash. The probability that 5 or more will be activated if the average number is 10 is almost 1. For an average number of 5, the probability that 5 or more will be activated is about 0.6. This means that if the average number is 5, in only about 60 percent of the flashes will a yes response be given. Also shown on the graph is the case when 6 or more rod cells must be activated, so that you can see how the curve depends on what the minimum number is. If you carried out such an experiment, the number of positive responses over the total number of trials would look very much like one of these curves as a function of the average number of rods activated. The curve is not a sharp step (as you would get if every flash contained exactly the

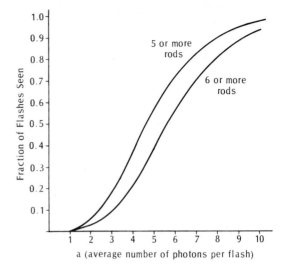

Figure 5.10

Fraction of flashes seen as a function of the number of photons per flash

same number of photons), but is a smooth, gradually rising curve, and this is the direct result of the Poisson distribution.

Problems

1. A person may keep his eyes dark-adapted if he works in a room filled with red light. Explain.

2. In an experiment using very weak flashes of light, it is found that the average number of photons per flash is 1.
   (a) What is the probability of getting less than 1 photon per flash?
   (b) What is the probability of getting more than 1 photon per flash?

3. An experiment is set up to determine the number of photons required to stimulate an impulse in the optic nerve of a cockroach. The nerve impulses are detected by an oscilloscope. The researcher beams flashes containing about 500 photons into

the insect's eye and estimates that in the average flash 5 photons strike the photoreceptors. To calculate the threshold of the eye he finds that he needs to know the percentage of the flashes in which two photons or less strike the retina. What is this percentage?

4. Approximately 4 percent of the photons directed at the pupil of the eye are absorbed by the rod cells. In an experiment in which 500 flashes of light containing 80 photons each, on the average, are presented to a subject, in how many flashes will the rod cells absorb one or two photons?

5. Suppose that, for a particular experimental subject, a flash of light is seen only if it consists of 2 or more photons. In a series of experiments, the average number of photons absorbed per flash by the rod cells was 3. Of 100 flashes presented to this subject, how many would he be expected to claim were not visible?

6. It is known that only 3 percent of the photons directed at the pupil of the eye are absorbed by the rod cells. In an experiment in which flashes of light contain, on the average, 100 photons, for what fraction of the flashes delivered do you expect no more than 1 photon is absorbed by the rod cells?

7. In light-scattering experiments, the intensity of the scattered light can be measured by counting the number of photoelectron pulses given off by a sensitive photomultiplier in some sampling time interval. In one experiment about 10^4 time intervals of 100 s each are sampled and the distribution of numbers in the time intervals studied. Given that the fraction of time intervals which contain 0 pulses is $f(0)$, the fraction with 1 pulse is $f(1)$ etc., show that these fractions obey the Poisson distribution given the researcher's data below:

No. of time intervals containing

zero pulses = 2231.0
1   pulse = 3347.0
2   pulses = 2507.0
3   pulses = 1254.0
4   pulses = 471.0

Chapter 6
Radiation Biophysics

6-1 Introduction

Studies of the effects of high energy radiation on biological molecules, cells and whole systems has been one of the most popular areas of biophysics for many years. As our society embarks on an energy program which is heavily based on nuclear reactors these studies are of great importance. In order to deal fairly and intelligently with the technological and political issues related to nuclear energy it is necessary that citizens become as knowledgeable of nuclear processes as they are of coal, oil, and more traditional fuels, especially with regard to ways in which the products of nuclear reactions affect biological systems.

6-2 Particles and Radiations of Significance

Radiation damage in biological systems can be due to either the high energy particles or the high energy photons produced in a nuclear process. The number of particles known to physics has grown extensively in the last three decades. However the number of those that occur in sufficient quantities to be a problem to the health of an earthbound population is fortunately quite small. The most important are undoubtedly the beta particle (electron) and the alpha particle (helium nucleus). Other particles, such as neutrons, could be a problem in some specialized research facilities or

in the case of the explosion of a nuclear weapon. A considerable flux of heavy metal ions occurs in outer space and could be a hazard for an extended space flight but fortunately for the vast majority, the earth's atmosphere offers sufficient protection from the major effects of these. The gamma rays, which are emitted during many naturally occurring radioactive decay processes, are high energy photons. X-rays, while originating from different sources, can be considered with gamma rays in discussing potential health hazards because they too are high energy photons.

There are several types of nuclear reactions. In the alpha particle (α) emission some unstable parent nucleus X of atomic number Z and atomic mass number A decays to yield the alpha particle of mass 4 and atomic number 2 plus a daughter nucleus of atomic number $Z-2$ and mass $A-4$. One or more gamma rays (γ) may be emitted virtually simultaneously to carry away excess energy. Symbolically,

$$ {}_{Z}^{A}X \rightarrow {}_{Z-2}^{A-4}Y + {}_{2}^{4}\alpha + \gamma \qquad Q_\alpha \text{ MeV} \qquad (6\text{-}1) $$

where Q_α represents the energy of the α particle. As a specific example consider the decay of polonium 210 to lead 206.

$$ {}_{84}^{210}\text{Po} \rightarrow {}_{82}^{206}\text{Pb} + {}_{2}^{4}\alpha + \gamma \qquad 5.3 \text{ MeV} \qquad (6\text{-}2) $$

The unit of energy is the MeV which is 10^6 eV where one electron volt (eV) is the energy acquired by one electron when accelerated through a potential difference of one volt. An eV has a value of 1.6×10^{-19} joule. The energy of the α particle is characteristic of the process.

In the case of beta particle (β) emission from some unstable parent nucleus the mass number is unchanged but the atomic number increases by one since β particle emission is equivalent to a nuclear neutron breaking down to a proton plus an electron. A gamma ray may also be emitted. Historically when β decays were first examined it seemed that there was a violation of the principle of conservation of mass-energy. To save the situation, physicists postulated that an additional particle with no charge and essentially zero mass was emitted at the same time as the β particle and, if detected, would restore the conservation of mass-energy. This "little neutral one" or neutrino has since been detected and possesses all the predicted characteristics. It is actually an anti-neutrino that is emitted. Neither the neutrino nor the anti-neutrino are known to have any biological significance.

The general equation for β decay is,

$$\,_{Z}^{A}X \rightarrow \,_{Z+1}^{A}Y + \beta^- + \bar{\nu} + \gamma \qquad Q_\beta \text{ MeV} \qquad (6\text{-}3)$$

Q_β is the total energy released. $\bar{\nu}$ represents the anti-neutrino. The β particles have a continuous distribution of energies from zero up to some maximum energy characteristic of the particular decay process. A specific example is,

$$\,_{82}^{214}\text{Pb} \rightarrow \,_{83}^{214}\text{Bi} + \beta^- + \bar{\nu} + \gamma \qquad 1.03 \text{ MeV} \quad (6\text{-}4)$$

The gamma rays emitted by a particular isotope have an energy spectrum that is characteristic of that isotope. This can be a particularly useful way of identifying trace amounts of, for example, heavy metals in a food sample, as it can be used with either naturally radioactive isotopes or with isotopes that have been excited by some process such as bombardment with x-rays or nuclear absorption of neutrons.

A few of the properties of the alpha, beta, and gamma radiations are listed in Table 6-1.

The two decay processes discussed, while quantitatively the most important γ-ray sources, are not the only γ-ray sources. A few artificially created radioactive nuclei decay via the emission of a positron (positive electron) and a γ. Other sources such as electron capture, isomeric transition, and spontaneous fission are discussed in more specialized texts.

Radioactive emission is a random process. Suppose, for example, the number of α particles emitted by a source are monitored for a certain time interval. It is found that the probability $P(n)$ of emission of a certain number, n, of particles obeys the Poisson distribution (see Chapter 4) for random events, that is

$$P(n) = a^n \frac{e^{-a}}{n!} \qquad (6\text{-}5)$$

Table 6-1
Properties of α, β, γ Radiations

Radiation	Rest Mass	Charge	Velocity $(c = 3 \times 10^8 \text{ ms}^{-1})$
α, alpha	6.68×10^{-27} kg	$+3.2 \times 10^{-19}$ C	variable, around 5% of c
β, beta	9.1×10^{-31} kg	-1.6×10^{-19} C	up to about 99% of c
γ, gamma	0	0	c

where a is the number of particles emitted on the average during the time interval.

6-3 Physical and Biological Half-Lives

Any specific radioactive isotope will decay according to the exponential function

$$N = N_o e^{-\lambda_p t} \quad \text{or} \quad \ln N/N_o = -\lambda_p t \quad (6\text{-}6)$$

where N_o is the activity (number of disintegrations per unit time) at time zero, N is the activity after an elapsed time, t, and λ_p is the decay constant for the isotope concerned. A graph of activity versus time gives a straight line if semi-logarithmic paper is used or the typical exponential curve on linear graph paper (see Figure 6-1).

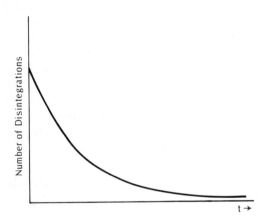

Figure 6.1

Exponential decay of radioactivity

Most references on radioactive materials list another parameter called the physical half-life of the isotope, T_p. This quantity is the time taken for the activity of a particular isotopic species to drop by a factor of two. Applying Equation (6-6) we see that

$$\ln \left(\frac{1}{2}\right) = -0.693 = -\lambda_p T_p \quad (6\text{-}7)$$

If for example, a quantity of some isotope with a half-life of three days were emitting 10,000 particles per second at some initial time, then three days later one would measure an activity of 5000 particles per second; in another three days 2500 particles per second and so on. A tremendous range in T_p exists for different radioactive elements: that of polonium (214) is less than a millisecond while that of uranium (238) is more than a billion years. Even different isotopes of the same element can have quite different half-lives. For example, T_p is 2.6 years for ^{22}Na and only 15 hours for ^{24}Na.

When an isotope enters some living system then biological processes, especially excretion, often act to decrease the amount of material present in the system. This biological removal of the isotope also progresses in an exponential manner. Thus

$$N = N_o e^{-\lambda_b t}$$

or

$$\ln \frac{N}{N_o} = -\lambda_b t \quad (6\text{-}8)$$

and

$$\ln \left(\frac{1}{2}\right) = -0.693 = -\lambda_b T_b \quad (6\text{-}9)$$

where λ_b and T_b are the decay constant and half-life associated with this biological process.

If the ingested isotope is a radioactive one then the physical and biological processes combine to give an overall, faster exponential decay. In this case

$$\ln \frac{N}{N_o} = \lambda_{\text{eff}} t \quad (6\text{-}10)$$

where λ_{eff} is the decay constant of the combined processes. Since

$$\lambda_{\text{eff}} = \lambda_p + \lambda_b = \frac{0.693}{T_p} + \frac{0.693}{T_b} \quad (6\text{-}11)$$

it is simple to combine the above equations and prove that the effective half-life T_{eff} is given by

$$T_{\text{eff}} = \frac{T_p T_b}{T_p + T_b} \qquad (6\text{-}12)$$

EXAMPLE

The physical half-life of an isotope is 20 days. After the isotope is administered to a patient, the measured activity is 25,000 counts s^{-1}. Five days later the activity is measured to be 10,000 counts s^{-1}. What is the biological half-life of this isotope?

$$\ln \frac{N}{N_o} = -\lambda_{\text{eff}} t$$

$$\ln \left(\frac{10^4}{2.5 \times 10^4} \right) = -\lambda_{\text{eff}} 5$$

$$\lambda_{\text{eff}} = 0.184 \text{ day}^{-1}$$

$$\ln \left(\frac{1}{2} \right) = -\lambda_{\text{eff}} T_{\text{eff}}$$

$$T_{\text{eff}} = -\frac{\ln 2}{-0.184} = 3.8 \text{ days}$$

$$T_{\text{eff}} = \frac{T_p T_b}{T_p + T_b}$$

$$3.8 = \frac{20 T_b}{20 + T_b}$$

$$T_b = 4.7 \text{ days}$$

6-4 Macroscopic Absorption of Radiation

In general as radiation passes through matter, events such as ionizations of atoms and molecules result in a loss of energy which eventually limits the range of the radiation. In general, the more massive the radiation the shorter the range. Thus, in any given material,

alpha particles will have the shortest range followed by beta particles and the gamma particles if the radiations are initially equally energetic. Also, the denser the medium being traversed, the shorter will be the range.

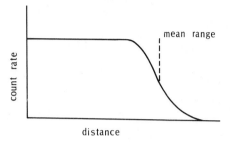

Figure 6.2

Typical range of α particles

Figure 6-2 indicates the relationship between measured count rate and distance as a detector is moved further and further away from a thin, flat, alpha source. For 5.3 MeV alpha particles, which are quite energetic, the mean range in air will be about 30 millimetres and in water or human tissue about 30 μm. These short ranges mean that externally originating alpha particles rarely constitute a health hazard. If the source is ingested, however, the α particles can be a very real hazard and in all cases any γ radiation accompanying the α decay may be a cause for concern.

As β particles traverse matter, they interact with orbital electrons and may have an abrupt change of direction for each interaction. Consequently the path will be quite tortuous and the integrated path length will be considerably greater than the actual range, which is the distance from the beginning to the end of the track. Within most materials of biological interest, the range will depend largely on the electron density which, in turn, is nearly proportional to the mass density. Thus beta particle ranges are frequently expressed in g cm^{-2} or mg cm^{-2}. For β energies (E) greater than

0.6 MeV the range is given approximately by the relationship,

$$Range = 0.542\,E - 0.133 \text{ g cm}^{-2}$$

$$\text{(with } E \text{ in MeV)} \quad (6\text{-}13)$$

EXAMPLE

What is the range of a 5.3 MeV electron?

$$Range = (0.542 \times 5.3) - 0.133$$

$$= 2.7 \text{ g cm}^{-2}$$

In normal air with density, $\rho = 1.29 \times 10^{-3}$ g cm^{-3}, the linear range will be $Range/\rho$.

$$\frac{Range}{\rho} = \frac{2.7 \text{ g cm}^{-2}}{1.29 \times 10^{-3} \text{ g cm}^{-3}}$$

$$= 2.1 \times 10^3 \text{ cm}$$

whereas in tissue of $\rho \approx 1.0$ g cm^{-3}, this figure drops to $\frac{2.7}{1} = 2.7$ cm.

In both cases the total path lengths would be about twice as great as the numbers calculated because of the scattering.

A graph of β intensity versus range, Figure 6-3, for decay of a particular radioisotope usually appears quite exponential. This is really a lucky accident because there is a range of β energies present. Absorption of monoenergetic β particles is not exponential.

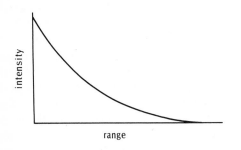

Figure 6.3

Intensity as a function of range for β particles

Most of the energy given up in biological systems by either α or β particles is used to create ion pairs in matter at an expenditure of about 34 eV per ion pair. Thus the 5.3 MeV electron can create about 1.56×10^5 ion pairs along its total path length of about 5.4 cm (2×2.7 cm, see preceding example) thus creating about 3×10^4 ion pairs cm^{-1} or 3 ion pairs μm^{-1}. Many living cells are a few micrometres long and thus several ion pairs could be created in the cell during traversal by such an electron.

Electromagnetic radiation such as gamma rays and x-rays are absorbed similarly. We will arbitrarily define gamma rays as quanta of wavelength 10^{-10} m (0.1 nm) or less with frequencies ν of 10^{18} s^{-1} or greater. Thus their minimum energy is,

$$E = h\nu$$

$$= 6.63 \times 10^{-34} \times 10^{18}$$

$$= 6.63 \times 10^{-16} \text{ J}$$

$$= \frac{6.63 \times 10^{-16} \text{ J}}{1.60 \times 10^{-19} \text{ J eV}^{-1}}$$

$$= 4.14 \times 10^3 \text{ eV}$$

$$= 4.14 \text{ keV}$$

X-rays typically have wavelengths between 200 nm and 2×10^{-3} nm. This corresponds to frequencies of about 10^{16} to 10^{21} s^{-1}. Thus x-rays and γ-rays overlap in wavelength and energy.

Some characteristics of absorption of γ-ray and x-ray quanta by materials are not all that different from UV or visible absorption which have been discussed earlier. For example, the overall absorption equation is

$$ln \frac{I}{I_o} = -\mu x \quad (6\text{-}14)$$

which has some similarity with the Beer-Lambert Equation (4-15). In Equation (6-14), I_o represents the incident intensity of monoenergetic γ-ray (or x-ray) photons, I is the intensity after passing a distance x through the material and μ is the linear absorption coefficient. The amount of absorption of γ-rays and x-rays is related to the density of the material, however, rather than the presence of certain chromophores. Thus

$$\mu = \mu_m \rho \qquad (6\text{-}15)$$

where ρ is the density of the material receiving the radiation. The coefficient μ_m, called the mass attenuation coefficient, is related to the probability that the photon will be absorbed. It depends upon the atomic composition of the substance and the energy of the incident photons. Combining Equations (6-14) and (6-15) one obtains

$$\ln\left(\frac{I}{I_o}\right) = -\mu_m \rho x \qquad (6\text{-}16)$$

Generally, the mass attenuation coefficient decreases with increasing photon energy. The 10 KeV x-rays used in most crystallographic studies have such a high value of μ_m in human tissue that if one is accidently exposed, almost all the photons are absorbed in the skin. The result is a painful burn which is slow to heal. Medical x-rays, however, are much higher in energy ($\approx 10^5$ eV). Consequently μ_m is much smaller and penetration through a body with relatively low absorption by tissue is quite practical. For a compound or mixture a weighted average of $\mu_m = \mu/\rho$ must be used for the wavelength concerned. Thus, for a sample 40 percent by weight of substance y and 60 percent of z the right hand side of Equation (6-16) would be

$$\left(\frac{0.4\,\mu_y}{\rho_y} + \frac{0.6\,\mu_z}{\rho_z}\right)\rho x$$

The absorption of high energy x-ray or γ-ray quanta occurs by three main processes; the photoelectric effect, the Compton effect and pair production. At photon energies less than 0.1 MeV, the photoelectric effect is predominant. A collision between this photon and

an electron of the material can result in the electron gaining a substantial kinetic energy. So much, in fact, that in some cases it can ionize neighbouring atoms or molecules. In the photoelectric effect, the photon gives up all its energy in the collision and hence disappears. In Compton scattering processes, which occur at energies greater than 0.1 MeV, the collision results in the electron assuming a kinetic energy which is only a fraction of the original photon energy (although the actual value of this kinetic energy is greater than was the case in the photoelectric effect). The photon is scattered and does not disappear. It is, however, reduced in energy and frequency.

Pair production occurs only at photon energies >1.02 MeV and since γ-rays of use in biophysics rarely have this much energy it is not too important. The result of pair production is that two particles, the electron and the positron, are created. Since the mass-energy (from $E = mc^2$) of these particles is 1.02 MeV, at least that much energy must be provided by the initial photon. If the photon energy exceeds 1.02 MeV, the excess energy is converted to kinetic energy of the particles.

6-5 Activity and Its Measurement

The amount of radioactivity of a material is normally measured in curies (Ci). One curie is 3.7×10^{10} radioactive disintegrations per second. A one-curie source represents a considerable hazard. Most experimental work involves microcurie or millicurie samples. A one-curie source of some isotope with a long half-life will, of course, retain a strength of approximately one curie for many years and should be treated with even more respect than a one-curie source with a short half-life which may decay to some negligible value in a year's time. The absolute determination of the total activity from a sample is a difficult procedure. Fortunately, relative activities of different samples are usually all that are required and are usually not too difficult. Consistency, however, is very important. Comparison between two samples should not be made unless they are of equal size, have identical histories, and are being measured with the same counter and the same geometry.

With most detection devices there is a "dead time" immediately following the detection of one event during which a second event cannot be detected. This dead time is quite short and for a low counting rate can be ignored. However at very high counting rates the presence of dead time can lead to appreciable underestimates of total activity. If the count rate from some sample, such as blood, were very high (say $10^5 \, s^{-1}$) then dead-time effects might be significant. If so, the total count obtained following division of the sample into two and dilution of each half back to its original volume with non-radioactive blood would give a rate greater than that obtained from the initial sample.

A considerable variety of detection devices can be employed depending on the type, energy, and intensity of the radiation to be measured. Geiger tubes, ionization chambers, proportional detectors, scintillation detectors, and solid state detectors are all used. The design and use of all of these is beyond the scope of this book.

The Geiger-Müller (G-M) tube consists of an outer cylindrical envelope, usually glass, with a fine tungsten wire along the central axis of the tube and a surrounding wire cylinder. The tube contains a small amount of gas. The central wire is maintained at a potential of about +1000 volts with respect to the wire cylinder. When a single ionizing particle passes through the tube it can produce a few gas ions which will initiate a discharge between the wire and the cylinder. An electronic counting system counts the number of these discharges. The G-M tube is useful whenever moderate or high-energy beta and gamma

particles are to be detected when precise knowledge of the energy is not important.

Scintillation detectors, which employ photomultiplier tubes to count the number of radiation induced flashes in certain crystals or liquids, give high efficiency for even low energy radiations and moderate resolution of the energy of the radiations. Much carbon-13 tracer work requires this type of detector.

Solid state detectors, such as lithium drifted germanium, give the highest energy resolution of all and thus can be particularly useful in work requiring element identification. Unfortunately these detectors are not as efficient in the detection process itself as are other types.

6-6 Units of Dose

When biological tissue is irradiated by x-rays or γ-rays only a small fraction of the incident photons is actually absorbed. For this reason it is more useful to use the absorbed energy (dose) rather than the energy passing through. The roëntgen (R) is such a unit of dose. It relates specifically to the absorption of radiation by air at 0°C and one atmosphere pressure. A dose of 1 R leads to the ionization of 2.08×10^{15} air molecules per cubic metre and hence produces that many ion pairs per cubic metre. This corresponds to about 1.6×10^{15} ion pairs per kilogram of air. Since it requires about 34 eV of energy for each ion pair formed then the amount of energy absorbed per kilogram of air exposed to 1 R will be,

$$E = 1.6 \times 10^{15} \times 34$$
$$= 5.4 \times 10^{16} \, eV \, kg^{-1}$$
$$= 8.6 \times 10^{-3} \, J \, kg^{-1}$$

Another unit of absorbed dose for any ionizing radiation is the rad.

$$1 \, rad = 0.01 \, J \, kg^{-1}$$
$$= 100 \, erg \, g^{-1}$$

The rad applies to any absorbing substance, not just air. For soft tissues exposed to γ-or x-rays the exposure in roëntgens will be approximately equal to the absorbed dose in rads.

The rem is a unit of dose equivalent equal to the dose in rads multiplied by a modifying factor, the RBE (as defined in the next section).

Most government regulations and recommendations oppose the presence of persons under 18 years of age in any work in radiation areas. A general recommendation is that the maximum whole-body dose equivalent should not exceed 5 rem per year from occupational sources. Thus at N years of age the maximum would normally be set at $5(N-18)$ rem. Any dose, no matter how small, is capable of causing some harm. There is great variability in susceptibility to radiation within the full range of biological systems with plants and lower forms of life being more resistant to damage than man and other animals. In man the whole-body dose which causes a 50 percent mortality rate in 30 days (known as the $LD_{50/30}$ dose) is around 300 to 500 rads.

6-7 Relative Biological Effectiveness (RBE)

The passage of three radiation species through a number of cells in some tissue is represented in Figure 6-4. Pairs of dots represent ion pair formation and therefore sites at which significant damage may (not will) occur. The density of dots also represents the LET (Linear Energy Transfer) of the radiations which is more or less inversely proportional to the range, since, if the LET is low, a particle of given energy will go further before it loses all its energy. A 5.4 MeV alpha particle has a LET of 110 keV μm^{-1} in water while a cobalt (60) γ-ray (about 1.3 MeV) has a LET of about 0.3 keV μm^{-1}.

The LET generally increases as the energy

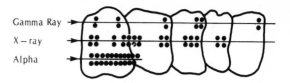

Figure 6.4
Passage of radiation through cells

decreases. For particles, this is because the electric field of a more slowly moving particle can act on some atomic electron for a greater length of time and therefore has a greater probability of driving the electron out and causing the ionization. Experiments have shown that ionizations are most likely to occur in clusters of about three with about 34 eV being given up for each ionization. You will recall that only about 3 eV is required to break an average covalent bond so the potential for damage at an ionization site is considerable.

The alpha particle has such a high LET that the cell traversed has a high probability of being destroyed whether only one hit or many hits is required. The alpha range is so limited that only very few cells are involved. At the other extreme a single gamma ray will pass through many cells causing damage in only a few and is capable of destroying only those where only a single hit is required. If a cell had several critical targets, then many gamma rays would have to pass through before damage could result. As a result of variations in the LET and the number of critical targets per cell there is, therefore, great variability in the RBE (Relative Biological Effectiveness) of different radiations.

Suppose a dose of 100 rads of beta particles was sufficient to destroy some biological system but 200 rads of x-rays were required for the same purpose. The betas would have twice the RBE of the x-rays. Very few generalizations can be made. Neutrons, for example, have a RBE of 20 for producing lens opacity but a RBE of only 1 for producing damage in many other organs. A RBE of 20 is meaningless unless we know to what the particular radiation is being compared. Unless otherwise stated, the standard damage is assumed to be that created by a 200 kV x-ray beam from which most of the long wavelength radiation has been removed by passage of the beam through a copper-aluminum filter.

6-8 Target Theory

Let us now look microscopically at the mechanism of radiation action. For the sake of discussion, imagine that a section of biological tissue, which is composed of individual cells, is being irradiated by a 0.5 MeV γ-ray source. With this energy, most absorption by the tissue will occur via the Compton effect. This process leads to the production of an energetic electron plus a scattered photon whose energy is several eV less than the incoming photon. This bundle of energy which is deposited is sufficient to ionize or fragment several molecules. The electron which receives this huge "kick" is immediately propelled out of its parent molecule at very high velocity leaving a positively charged ion behind it. This is called the primary ionization. The electron, which is called a secondary electron, is so energetic that it is capable of ionizing or fragmenting several other molecules. In some books this fast secondary electron is called a delta ray although it really behaves exactly as a β-ray (also a fast electron).

Because of its high LET, the delta ray quickly yields its energy to nearby molecules, causing, depending on its energy, several secondary ionizations. All this action occurs within a small volume which is less than 1 nm in diameter around the primary ionization. This small volume is sometimes called an inactivating event. The distribution of inactivating events throughout the irradiated tissue is *completely random*; there is no way of predicting the exact spot where one will occur. The reason for it being termed an inactivating event is that the primary ionization plus the secondary ionizations which follow it may, if the event happens to occur at just the right spot, inactivate or kill one of the cells of the tissue. Thus an inactivating event has only the potential to inactivate: cell inactivation as we shall see shortly is reasonably rare, at least for ordinary doses of radiation.

The number of inactivating events produced in the tissue is directly proportional to the radiation dose it receives. If I represents the number of inactivating events per cubic centimetre then

$$I = 6.1 \times 10^{11} D \qquad (6\text{-}17)$$

where D is the dose in rads.

Since the inactivating events are distributed randomly through the tissue we can use the Poisson distribution to determine the probability that the cells will receive 0, 1, 2, or more events. This does not tell us how many cells will die since many of the inactivating events occur in parts of cells which are either unimportant or else parts which are easily replaced by normal metabolism. When applied in this situation the Poisson distribution becomes

$$P(n) = \frac{e^{-IV}(IV)^n}{n!} \qquad (6\text{-}18)$$

where V is the volume of the biological cell. The product, IV, is the average number of inactivating events per cell.

EXAMPLE

A tissue containing cells of dimensions $(10 \ \mu m) \times (5 \ \mu m) \times (5 \ \mu m)$ is irradiated to a dose of 0.01 rads. The average number of events per cell is

$$VI = 10 \times 10^{-4} \times (5 \times 10^{-4})^2 \times 6.11 \times 10^{11}$$
$$\times 0.01$$
$$= 1.53$$

Thus the fraction of cells which do not receive any events is, using Equation (6-18).

$$P(0) = e^{-1.53}$$
$$= 0.22 \quad \text{or} \quad 22 \text{ percent}$$

The fraction receiving 1, 2, 3 events are

$$P(1) = \frac{e^{-1.53} 1.53}{1}$$
$$= 0.34 \quad \text{or} \quad 34 \text{ percent}$$

$$P(2) = \frac{e^{-1.53} (1.53)^2}{2}$$
$$= 0.26 \quad \text{or} \quad 26 \text{ percent}$$

$$P(3) = \frac{e^{-1.53} (1.53)^3}{3 \times 2}$$
$$= 0.13 \quad \text{or} \quad 13 \text{ percent}$$

etc.

The probabilities found by using Equation (6-18) allow one to find the fraction of cells receiving a certain number of inactivating events, but provide no information regarding the fraction of cells which would be killed by a certain dose. It is often useful, however, to describe another volume, called a sensitive volume, V_s, which, when hit by an inactivating event, will indeed lead to cell death. The basic premise here is that if a cell's sensitive volume is hit, the cell dies; if it is missed, the cell is unaffected. This sensitive volume of a cell is much smaller than the cellular volume and has been shown to be the volume of the cell's DNA. If each cell contains one sensitive volume, then it is easy to see that the probability that cells will survive corresponds to the probability that their sensitive volumes are missed. If N_o is the original number of cells and N is the number surviving the radiation then the probability of survival is simply N/N_o. Thus

$$P(0) = \frac{N}{N_o} = e^{-IV_s} \qquad (6-19)$$

or

$$ln \frac{N}{N_o} = -IV_s$$

One of the interesting uses of this equation is the determination of a cell's sensitive volume. If the surviving fraction of cells is plotted on a semi-logarithmic graph against I (or D) then the sensitive volume (V_s) can be obtained from the slope of the line which is the best fit to the data points. Note also that if the radiation dose is such as to make the average number of events per sensitive volume (IV_s) equal to one then

$$\frac{N}{N_o} = e^{-1}$$
$$= 0.37$$

This dose, which is often called the D_{37} is demonstrated in Figure 6-5.

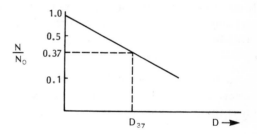

Figure 6.5
Origin of the D_{37} dose

If the system being inactivated is an enzyme molecule of density ρ then the molecular weight of this protein will be

$$M = V_s \rho N_A$$

where N_A is Avogadro's number. Typical values of ρ for protein lie around 1.3 g cm^{-3}. Using this method, molecular weights of 30,000 and 230,000 have been found for ribonuclease and the enzymatic portion of myosin. These values are in good agreement with those obtained by other methods. However, this method is not always successful and quite poor estimates of molecular weight have been found for some materials principally because the volume, V_s, for the enzyme is often considerably different from its true volume V.

6-9 Multitarget Theory

In many circumstances the system being irradiated may have several targets, all of which must be hit before inactivation occurs. As an example, consider again the biological cell. If there are p targets in the cell and all must be hit for the cell to die, it is possible to evaluate a new survival equation. The probability of hitting one of the targets is one minus the probability of a miss; that is

$$1 - P(0)$$

or

$$1 - e^{-IV_s}$$

If there are p targets, the probability of them all being hit is the product of all the individual probabilities or

$$(1 - e^{-IV_s})^p$$

Since this is the probability of hitting all p targets, the probability of them *not* all being hit is the probability of survival N/N_o. Thus

$$\frac{N}{N_o} = 1 - [1 - e^{-IV_s}]^p \qquad (6\text{-}21)$$

EXAMPLE

For many biological cells there are two targets. Sketch a graph of the surviving fraction N/N_o as a function of dose.

(a) For low doses I is very small (i.e. approaches zero). Therefore

$$e^{-IV_s} \approx e^{-0} \approx 1$$

and Equation (6-21) becomes

$$\frac{N}{N_o} = 1 - (1 - 1)^2 = 1$$

(b) For large doses

$$\frac{N}{N_o} = 1 - [1 - e^{-IV_s}]^2$$

$$= 1 - (1 - 2e^{-IV_s} + e^{-2IV_s})$$

Since e^{-2IV_s} is very small relative to $2e^{-IV_s}$, it can be neglected and

$$\frac{N}{N_o} \approx 2e^{-IV_s}$$

(c)

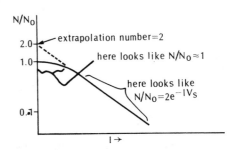

Note that the extrapolation number is equal to the number of targets.

Target theory has been used for many years to evaluate the size of DNA in cells, the number of targets, the size of enzymes, the size of virus genetic material and many other quantities. Often it is very successful. Often, because of special problems, some of which will be discussed in the next section, it gives poor results. So it must be applied cautiously and with as many experimental checks as possible.

6-10 Action of Radiation at the Molecular Level

Damage to biological systems occurs because crucial molecules within the individual cell are altered directly or indirectly as radiation passes through a living cell. A bond in a DNA molecule may be broken due to a direct ionization by radiation, or free radicals produced by ionization of water molecules may attack the DNA and cause, indirect damage. In some species, if only one strand of the DNA is altered and the cell is in a resting stage a special mechanism can cleave out and replace the damaged section. Many species do not possess this repair mechanism and even if they did it might not have time to operate if the cell is in rapid growth. Both the indirect effect and the repair mechanisms can produce uncertainty in experiments based on target theory. Robert Haynes and his colleagues are conducting experiments to clarify the importance of repair in survival curves such as the one shown in the previous example.

The nucleic acids are not, of course, the only molecular types subject to damage. Damage to protein molecules can also seriously impair cell function. Scission, or breaking of the primary chain, or cross-linking between two parts of the primary chain is often sufficient to impair the action of an enzyme. Even the knocking off of a group as small as an NH_3 group may be serious especially if it is located in the active site of an enzyme (see Chapter 8).

Some of the indirect (solvent) type effects are brought about by processes, initiated by the ionizing radiation, such as,

$$H_2O \rightarrow HO\cdot + H\cdot$$

$$H_2O \rightarrow H_2O^+ \rightarrow H^+ + HO\cdot$$
$$+ e^- \searrow HO^+ + H\cdot$$

$$H_2O + e^- \rightarrow H_2O^- \rightarrow OH\cdot + H\cdot^-$$
$$\searrow OH^- + H\cdot$$

The dots signify free radicals which are extremely reactive. They may alter proteins and nucleic acids directly or recombine to form compounds such as H_3O which are chemically very reactive.

Problems

1. Complete the reactions and identify all components
   (a) $^{226}_{88}Ra \rightarrow ^{222}_{86}Rn +$
   (b) $^{210}_{82}Pb \rightarrow ^{210}_{83} \quad +$
   (c) $^{4}_{2}He + ^{9}_{4}Be \rightarrow ^{12}_{6}C +$

2. What is the half-life of a radioactive substance for which the decay constant is $2 \times 10^{-6}\,s^{-1}$?

3. The half-life of ^{28}Al is 2.3 minutes and of ^{24}Na is 15 hours. If a source initially contains 1,000 times as many aluminum as sodium atoms, what will be the ratio of the number of aluminum to sodium atoms remaining after 10 minutes? After 30 minutes?

4. Ingested ^{131}I is regularly used in clinics to destroy overactive or diseased thyroid tissue (the so-called radioactive cocktail). Because of your expert background in biophysical matters you've been asked to determine the biological half-life of the isotope in a particular patient's thyroid gland. One day a counter positioned near the gland reads 10^4 counts per second. Forty-eight hours later you measure a count rate of 6000 counts per second. What do you calculate the biological half-life of ^{131}I in the thyroid to be? (The physical half-life of ^{131}I is 8.1 days.)

5. The physical half-life (T_p) of a radioactive isotope is 15 hours. The biological half-life is 30 hours. If after ingestion of this isotope, the observed count rate of a blood sample is 4×10^4 counts per second, how long will it take for the count rate to fall to 4×10^3 counts per second?

6. An isotope with a half-life of 15 days is ingested by a cow. A measurement of the activity of a sample of the cow's blood yielded a count rate of 7500 counts per second. Twenty-four hours later a similar check yielded a count rate of 2500 counts

per second. What is the biological half-life of this isotope?

7. In the radioactive decay of ^{116}In many β particles of energy approximately 1 MeV are given off. What would be the approximate range of these βs in human tissue? Through how many "typical" cells of thickness 20 micrometres could the βs penetrate?

8. An x-ray unit operates at 50,000 volts. What is the maximum energy of x-rays that can emerge from the unit? What will be the wavelength of these x-rays? If linear absorption coefficient in human tissue is approximately $7.5\,cm^{-1}$ through what thickness of tissue will a beam of x-rays pass before the intensity is decreased to one-half?

9. The mean lethal radiation dose in man is about 500 rads. How much energy is deposited per gram of tissue by a dose of this magnitude?

10. A dose of 500 rads of x-rays is required to destroy 63 percent of a bacterial population. 100 rads of β particle radiation destroys a similar amount. What is the RBE of the β particle as compared to that of the x-ray?

11. A piece of bone of thickness 3.0 cm and density $3.0\,g\,cm^{-3}$ is placed between a γ-source and detector. The count rate drops from 1000 counts per minute to 400 counts per minute. What is the mass attenuation coefficient of the bone?

12. The count rate from a γ-source is measured to be 1000 counts per minute. When 10.0 cm of lead is placed between the source and detector, the count rate is reduced to 600 counts per minute. What additional thickness of lead would have reduced the count rate to 500 counts per minute?

13. After a group of cells have been radiated by a dose of 200 rads you find that the fraction surviving is 0.57. If there is only one target per cell, what is its sensitive volume? (Assume 6.1×10^{11} events per cm^3 per rad.)

14. Mammalian cells each having a volume of $3.6 \times 10^{-11}\,cm^3$ are irradiated by γ-rays. It is found that 1.0×10^{11} inactivating events are produced per cm^3 of the cells. What fraction of the cells experienced no inactivating events?

15. There are about 6.1×10^{11} inactivating events per cm^3 per rad. How many inactivating events per cm^3 are produced by a typical medical x-ray of dose 10 millirads?

16. If the volume of a cell is considered to be about $2 \times 10^{-11}\,cm^3$, what fraction of the cells of the person getting the x-ray in the previous question would be hit?

17. Hela cells are a particular strain of malignant human cells which can be cultured quite easily. The following table shows the surviving fraction of hela cells as a function of x-ray dose.

Surviving Fraction	Dose (rads)
0.82	60
0.72	85
0.50	100
0.45	150
0.31	200
0.18	250
0.11	300
0.06	400
0.01	500

(You will need semi-log paper for this question. There are about 6.1×10^{11} inactivating events per cm^3 per rad).

(a) Find the number of sensitive targets in the hela cell.

(b) Find the sensitive volume (volume of the DNA) of the cell.

Chapter 7
Animal Mechanics

7-1 Introduction

The form and lifestyle of an animal is determined by the physical forces which maintain supreme control in its particular environment. For mammals like man, the most important force is gravity for it determines our size and it restricts our movements. The human body is an exceedingly complex entity. It moves with a smoothness and a coordination that no machine can emulate yet the motion that it can perform follows the familiar laws of classical mechanics. It is the purpose of this chapter to show how the concepts of classical mechanics can be used to analyse an animal's motion, its strength, its size, and the properties of its component material.

7-2 Force—A Vector

Physical quantities can usually be classified into two groups; scalars, which are characterized by their magnitude as, for example, time or temperature, and vectors which require both a magnitude and a direction for their complete definition. Force, be it the force of friction or the force of air resistance or the force arising from the gravitational attraction of the earth, is a vector. Often a vector is represented schematically by an arrow ↗ where the length of the arrow is proportional to the magnitude of the vector and the arrow head represents the direction of the vector.

Sometimes a body has many forces acting on it simultaneously. The effect of several forces can be determined by combining the force vectors according to the parallelogram law of addition. Consider Figure 7-1. $\vec{F}_1$ and $\vec{F}_2$ are two forces acting at the point A, their magnitudes and directions specified by the length and orientation of the arrows. The resultant force $\vec{F} = \vec{F}_1 + \vec{F}_2$ is simply the diagonal AB of the parallelogram as indicated in the diagram. Note, $\vec{F}_1 + \vec{F}_2$ is a vector sum not an algebraic sum. If more than two forces are present simply find the resultant of any two and combine this with the third and so on.

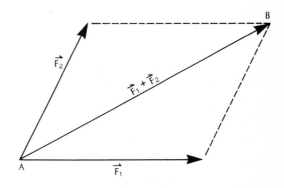

Figure 7.1
Parallelogram law of addition for vectors

The combination of forces to find the resultant force can also be accomplished by a process called resolution of vectors into components. A two-dimensional vector is uniquely determined by its components along two orthogonal axes (x, y). The component of a force along an axis is the magnitude of the force multiplied by the cosine of the angle between the force and the axis. Each vector is resolved into its x-component and its y-component. The problem is thus reduced to two one-dimensional problems where the vector components can be added directly (algebraically) to give the vector components of the resultant force. Figure 7-2 shows two forces $\vec{F}_1$ and $\vec{F}_2$ acting at an angle θ to each other.

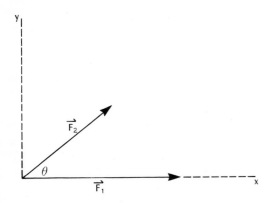

Figure 7.2

Two forces acting at an angle θ

An orthogonal coordinate system has been chosen such that $\vec{F}_1$ lies along the x-axis and $\vec{F}_2$ makes an angle θ with the x-axis. The x-component of $\vec{F}_1$ is

$$F_{1x} = F_1 \cos 0°$$
$$= F_1$$

and the x-component of $\vec{F}_2$ is

$$F_{2x} = F_2 \cos \theta$$

The y-components of $\vec{F}_1$ and $\vec{F}_2$ are

$$F_{1y} = F_1 \cos 90°$$
$$= 0$$

$$F_{2y} = F_2 \cos (90 - \theta)$$
$$= F_2 \sin \theta$$

The resultant x-component is

$$F_x = F_{1x} + F_{2x}$$
$$= F_1 + F_2 \cos \theta$$

and the resultant y-component is

$$F_y = F_{1y} + F_{2y}$$
$$= F_2 \sin \theta$$

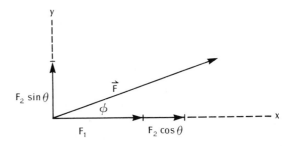

Figure 7.3

Determination of the resultant force using components

Figure 7-3 shows the x- and y-components of the resultant force. The magnitude of the resultant force is determined by the Pythagorean theorem to be

$$F = \sqrt{F^2}$$
$$= \sqrt{F_x^2 + F_y^2}$$
$$= \sqrt{F_1^2 + F_2^2 + 2F_1F_2 \cos \theta}$$

and its direction can be determined from

$$\tan \phi = \frac{F_y}{F_x}$$
$$= \frac{F_2 \sin \theta}{F_1 + F_2 \cos \theta}$$

EXAMPLE

The tension T in the quadriceps tendon as it passes over the knee cap is 1500 N. What is the magnitude and direction of the resultant force $\vec{F}$ exerted on the femur by the cap for the configuration shown in the diagram.

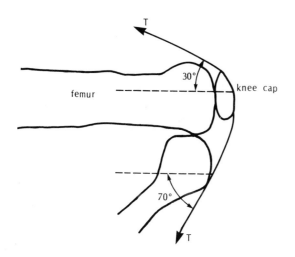

$$\sum F_x = T \cos 30° + T \cos 70°$$
$$= 1500\,(0.86 + 0.34)$$
$$= 1812 \text{ N}$$
$$\sum F_y = T \sin 70° - T \sin 30°$$
$$= 1500\,(0.94 + 0.50)$$
$$= 660 \text{ N}$$

The resultant force has a magnitude of

$$F = \sqrt{1812^2 + 660^2}$$
$$= 1930 \text{ N}$$

and a direction given by

$$\tan \theta = \frac{660}{1812}$$
$$= 0.36$$
$$\theta = 20°$$

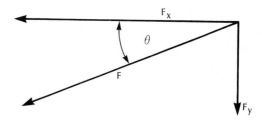

Draw a diagram showing the vertical and horizontal components of the two tension forces.

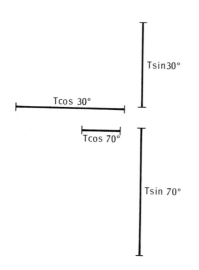

7-3 Friction

Friction is an example of a force which plays an important role in animal mechanics. It can aid or impede motion. The friction between the rubber soles of running shoes and the track gives a sprinter the traction required for a fast start. On the other hand, excess friction in joints can be a severe impediment to the proper operation of the joint.

Frictional forces oppose the movement of one surface over another. If the two surfaces are at rest then a certain force will have to be applied to bring one to the point of moving

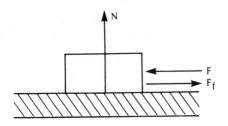

Figure 7.4
Friction

(Figure 7-4). This limiting force F is proportional to the normal force N at right angles to the two surfaces, that is

$$F = \mu_s N \qquad (7-1)$$

where μ_s is termed the coefficient of static friction. The frictional force $\vec{F}_f = -\vec{F}$. Once in motion the frictional force decreases and is characterized by a coefficient of kinetic friction μ_k where usually $\mu_k < \mu_s$. The coefficients of friction depend primarily on the nature of the surfaces in contact. For rough surfaces they are large, for smooth surfaces they are small. The above equation is not exactly valid for complicated surfaces like those of plastics or biomaterials but is a reasonable approximation.

If frictional forces are to be overcome energy must be expended usually in the form of heat. In joints this is damaging because proteins are easily denatured by heat. Also if the frictional forces are very large the muscle can easily be overextended to the point of damage trying to force joint movement. Lubrication can reduce friction, for example, moving parts in machinery are oiled. Human joints contain a remarkable lubricating liquid called synovial fluid. The properties of synovial fluid are discussed in Chapter 9.

EXAMPLE

Sniff has created a new sport: "roof-skiing." He slides down the snow-covered roof of his dog house. Assuming that he has forgotten to wax his skis and that the snow is sticky what is the coefficient of static friction between his skis and the snow if he slides down the roof without giving himself an initial push? The angle of inclination of the roof is 30° to the horizontal. Draw a force diagram.

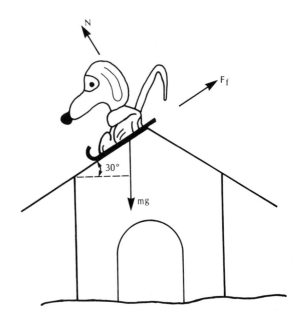

The force of friction $\vec{F}_f$ acts directly up the plane. A component of the gravitational force mg acts down the plane. This component has the value

$$F = mg \sin 30°$$

The normal reaction force acts perpendicular to the plane and is equal to the other component of the gravitational force

$$N = mg \cos 30°$$

If Sniff is to ski down the roof

$$\vec{F} \geq \vec{F}_f$$
$$F_f = \mu_s N$$
$$= \mu_s mg \cos 30°$$

Therefore

$$mg \sin 30° \geq \mu_s mg \cos 30°$$

$$\mu_s \leq \tan 30°$$

7-4 Equilibrium

If the resultant force acting on a body is zero, the body is said to be in a state of translational equilibrium which means that it is either at rest or moving with a constant velocity. The condition for translational equilibrium is stated simply as follows. The vector sum of all forces acting on the body must be zero.

$$\sum_n \vec{F}_n = 0 \qquad (7\text{-}2)$$

Frequently, there are forces exerted which tend to make a body rotate as in closing a door or dialling a telephone. A measure of the effectiveness of a force in causing a body to rotate is termed the turning moment (or torque) of the force and is the product of the magnitude of the force and the perpendicular distance from the axis of rotation of the body to the line of action of the force (Figure 7-5). A force cannot exert a moment about a point through which it passes. If a body is in a state of rotational equilibrium then the sum of the moments of all forces tending to make that body rotate about any particular point must be zero. Counter-clockwise moments are considered positive, clockwise moments negative.

$$\sum_n M_n = 0 \qquad (7\text{-}3)$$

For an object to be in a state of complete static equilibrium both Equation (7-2) and Equation (7-3) must be satisfied simultaneously.

In order to investigate the equilibrium properties of an extended body it is convenient to know the position of the centre of mass of the body. The centre of mass of an object is the point at which the total mass can be considered to be located for translational motion. The position of the centre of mass depends on the physical properties of the body and can be localized inside or outside the body. For example, the centre of mass of a perfect sphere is at the centre of the sphere while the centre of mass of a human body, seated with his arms extended in the forward direction, is just slightly in front of his navel. The determination of the exact position of the centre of mass can be a rather laborious process unless the object has a high degree of symmetry.

Using the concepts of static equilibrium it is easy to determine the forces present in many physical systems as the following example demonstrates.

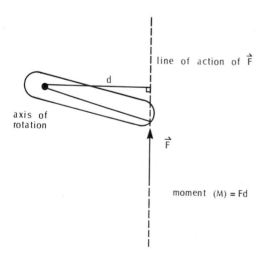

line of action of $\vec{F}$

d

axis of
rotation

$\vec{F}$

moment (M) = Fd

Figure 7.5
Definition of the moment of a force

EXAMPLE

Calculate the tension in the deltoid muscle of the arm arising from elevation of the arm.*

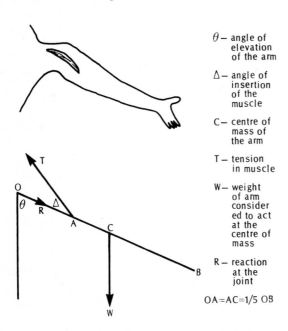

θ — angle of elevation of the arm

Δ — angle of insertion of the muscle

C — centre of mass of the arm

T — tension in muscle

W — weight of arm considered to act at the centre of mass

R — reaction at the joint

$OA = AC = 1/5 \ OB$

*L. A. Strait, V. T. Inman and H. J. Ralston, "Sample Illustrations of Physical Principles Selected From Physiology and Medicine", Am. J. Phys. 15 (1947), 375.

The arm is at rest in its elevated position so it is in a state of rotational equilibrium. Any point may be chosen about which the moments of the forces can be evaluated. Let us use the point O.

$$T(OA) \sin \Delta - W(OC) \sin \theta = 0$$

$$T = W \frac{\sin \theta}{\sin \Delta} \frac{OC}{OA}$$

$$= W \frac{\sin \theta}{\sin \Delta} \frac{2}{5} \cdot \frac{5}{1}$$

$$= 2W \frac{\sin \theta}{\sin \Delta}$$

7-5 Motion

Almost all animals are capable of motion. It is necessary for survival, to search for food, to escape from danger. Some animals are capable of great speed (the cheetah can sprint at $110 \ km \ hr^{-1}$). Others, while slower moving, possess greater endurance (the horse can maintain a speed of $24 \ km \ hr^{-1}$ for a distance of 55 km). Kinematics is the study of motion. General movement consists of linear or translational components and rotational components. For example: during athletic activities the human body executes linear and rotational motion simultaneously and the two must be coordinated for the best result; the swivelling shoulder blades of the horse and cheetah add many centimetres to their strides. We shall see in this section how to analyse linear and rotational motion.

Linear motion is characterized by the variables displacement ($\vec{s}$), velocity ($\vec{u}, \vec{v}$), acceleration ($\vec{a}$), time (t) where $\vec{s}, \vec{u}, \vec{v}, \vec{a}$ are vector quantities. For convenience let us consider motion in one dimension only so we need consider only the magnitude of the quantity. This is not a serious constraint because of course any general three dimensional motion can be analysed in terms of the three independent directions or in terms of vector addition. Velocity is the rate of change of displacement with time and is expressed as

$$v = \frac{ds}{dt} \qquad (7\text{-}4)$$

Acceleration is the rate of change of velocity with time and is expressed as

$$a = \frac{dv}{dt} \qquad (7\text{-}5)$$

or by using Equation (7-4) as

$$a = \frac{d^2 s}{dt^2} \qquad (7\text{-}6)$$

For the case of uniform acceleration various simple relationships exist linking the parameters s, u, v, a, t. They can be derived as follows. Equation (7-5) can be written in its integral form as

$$\int_u^v dv = a \int_0^t dt$$

where u designates the initial velocity and v the final velocity. Carrying out the integration yields

$$v - u = at$$

or

$$v = u + at \qquad (7\text{-}7)$$

Equation (7-4) can be written in its integral form as

$$\int_0^s ds = \int_0^t v\, dt$$

or

$$\int_0^s ds = \int_0^t (u + at)\, dt$$

By integration

$$s = ut + \tfrac{1}{2}at^2 \qquad (7\text{-}8)$$

Eliminating either t or a between Equations (7-7) and (7-8) leads to two other equations

$$v^2 = u^2 + 2as \qquad (7\text{-}9)$$

$$s = \left(\frac{u+v}{2}\right)t \qquad (7\text{-}10)$$

Any problem involving linear motion with constant acceleration can be tackled using Equations (7-7), (7-8), (7-9) and (7-10). Consider the following examples.

EXAMPLE

A cheetah can sprint at a velocity of $110\ \text{km hr}^{-1}$. The best man is capable of is a velocity of $35\ \text{km hr}^{-1}$. A man and a cheetah are 0.4 km apart. Neglecting any initial acceleration and assuming that both man and cheetah are running at their top speed how long does it take the cheetah to overtake the man?

It is always a good idea in problems of this nature to sketch a diagram showing the initial and final conditions.

The man covers the distance s in t hr, the time it takes for the cheetah to move a distance $(s+0.4)$. Using Equation (7-10)

$$\text{man} \quad s = \frac{(35+35)t}{2}$$

$$= 35t$$

$$\text{cheetah} \quad (s+0.4) = \frac{(110+110)t}{2}$$

or

$$t = \frac{s+0.4}{110}$$

$$= \frac{35t+0.4}{110}$$

$$110t - 35t = 0.4$$

$$75t = 0.4$$

$$t = 5.3 \times 10^{-3}\ \text{hr} \quad \text{or} \quad 19\ s$$

EXAMPLE

A honey bee is visiting a clover patch. Its average speed between flowers is $2.00\ \text{m s}^{-1}$. Starting at one flower it travels north for $0.20\ \text{s}$ to another flower, then east for $0.30\ \text{s}$ to another flower, and finally south-east for $0.10\ \text{s}$ to another flower. What is the bee's final displacement and bearing from the first flower?

Account must be taken of the vector nature of the parameters. However, before attempting to solve this example it is best to lay out the problem in a diagram. This is the best mode of attack for almost any problem you will encounter.

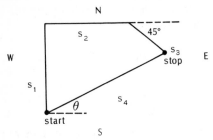

Using Equation (7-8)

$$s_1 = ut_1 = 2(0.20) = 0.40 \text{ m}$$

$$s_2 = ut_2 = 2(0.30) = 0.60 \text{ m}$$

$$s_3 = ut_3 = 2(0.10) = 0.20 \text{ m}$$

The east-west component of

$$s_4 = s_2 + s_3 \cos 45° = 0.74 \text{ m}$$

The north-south component of

$$s_4 = s_1 - s_3 \sin 45° = 0.26 \text{ m}$$

The bee's displacement from the start is given by

$$s_4 = \sqrt{0.74^2 + 0.26^2}$$

$$= 0.75 \text{ m}$$

The bee's bearing from the start is given by

$$\theta = \tan^{-1} \frac{0.26}{0.74}$$

$$= 19° \text{ north of east}$$

Rotational motion is characterized by the variables θ (angular displacement in radians), ω (angular velocity in radians s^{-1}), α (angular acceleration in radians s^{-2}). If there is uniform acceleration it is easy to show that the same expressions as for linear motion are valid with angular measure exchanged for linear measure. That is

$$\theta = \frac{(\omega_o + \omega)t}{2}$$

$$\theta = \omega_o t + \tfrac{1}{2}\alpha t^2 \qquad (7\text{-}11)$$

$$\omega = \omega_o + \alpha t$$

$$\omega^2 = \omega_o^2 + 2\alpha\theta$$

EXAMPLE

A figure skater in a spin increases her angular velocity from $\omega_o = 0$ to $\omega = 6$ radians s^{-1} in 0.5 s under constant acceleration conditions. (a) What is her angular acceleration? (b) What angle does she rotate through during the acceleration?

(a)

$$\omega = \omega_o + \alpha t$$

$$\alpha = \frac{\omega - \omega_o}{t}$$

$$= \frac{6 - 0}{0.5}$$

$$= 12 \text{ radians s}^{-2}$$

(b)

$$\theta = \frac{(\omega_o + \omega)t}{2}$$

$$= \frac{(0 + 6)}{2}(0.5)$$

$$= 1.5 \text{ radians}$$

7-6   Force and Motion—Dynamics

We have discussed force and motion separately in some detail. There remains however the task of linking the two concepts together. Dynamics is the study of motion with respect to the forces that give rise to it. Newton stated the three fundamental laws of dynamics.

(i) A body remains in a state of rest or uniform motion in a straight line unless acted on by an unbalanced force.

(ii) An unbalanced force ($\vec{F}$) acting on a body results in an acceleration ($\vec{a}$) in the direction of the force. The magnitude of the acceleration is directly proportional to the magnitude of the force and inversely proportional to the mass (m) of the body.

$$\vec{F} = m\vec{a} \qquad (7\text{-}12)$$

(iii) For every action force there is an equal and opposite reaction force.

Therefore, according to Newton's Laws, in linear motion a body accelerates only if there is an external unbalanced force acting. If all external forces are removed the body moves at a constant velocity (which can be zero, i.e. the body is at rest) and we have the case of translational equilibrium which was investigated in Section 7-4.

Let's look at rotational motion in some detail, in particular uniform circular motion which is very important, for instance in the action of a centrifuge. An object executing uniform circular motion appears to be moving with a constant velocity and indeed it is as far as the magnitude of the velocity is concerned. But remember that velocity is a vector, and in circular motion the direction of that velocity vector is changing as a function of time, hence there is a non-zero acceleration and by Newton's second law a force acting on the object. It is called a central or centripetal force and is given by the expression

$$F = \frac{mv^2}{r} \qquad (7\text{-}13)$$

or

$$F = m\omega^2 r$$

where m is the mass of the object, r is the radius of the circle transcribed, v is the linear velocity of the object, ω is the angular velocity of the object. The quantity v^2/r or $\omega^2 r$ in Equation (7-13) is the central acceleration.

The rotational motion of a body is further characterized by a quantity called its angular momentum, $I\omega$, where I is the moment of inertia of the body for rotation about a particular axis and ω is the angular velocity of the body about that axis. The angular momentum of a body is a constant in the absence of unbalanced external turning moments. Of course if there are no unbalanced external turning moments there is no angular acceleration and we have the case of rotational equilibrium.

The moment of inertia of a body is an important parameter in all physical systems. The moment of inertia of an object in a given situation measures the reluctance of the object to rotate in that position. It depends on the mass of the object and what is more important it depends on the distribution of the mass about the axis of rotation. The moment of inertia of a body is given by

$$I = \sum_i m_i r_i^2 \qquad (7\text{-}14)$$

where m_i is the mass of the individual particles, r_i is the distance of each particle from the axis of rotation, $\sum$ means to sum the contribution from all particles in the body, Figure 7-6. In the case of a continuous object the sum in Equation (7-14) becomes an integral. The value of I can be calculated from Equation (7-14) for objects of simple geometry.

The moments of inertia for an object can have different values for rotation about different axes. The value of the moment of inertia about a particular axis can change if different mass configurations are possible, as is the case with the human body. Table 7-1 lists measured moments of inertia for the human body about its three principal axes of rotation.

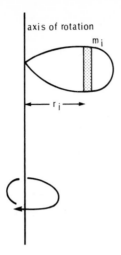

Figure 7.6
Definition of the terms in Eq. (7-14) for any object rotating about an axis

Table 7-1
Moments of Inertia for Human Body

axis	moment of inertia
vertical	$I \simeq 1$ kg m^2
transverse horizontal	$I \simeq 11$ kg m^2
transverse medial	$I \simeq 11$ kg m^2

To demonstrate the application of kinematics and dynamics in physical systems consider the following discussions on the everyday processes of walking and jumping as executed by the human body.

WALKING

The movement of the foot during the walking process is analogous to the simple harmonic motion (Appendix 3) executed by a simple pendulum. Using the formalism of harmonic motion the velocities and accelerations of the foot can be calculated. The speed with which the foot moves during walking is really quite surprising when compared to the forward velocity of the body as a whole. The result of course is the common malady known as sore feet.

The motion of a simple pendulum is familiar to all, Figure 7-7. The pendulum swings about an equilibrium position, where the amplitude of the swing is designated by x_o. The time for one complete cycle is known as the period T. The pendulum comes momentarily to rest at the extremities of the swing, and has its maximum velocity when it crosses the

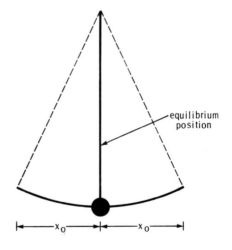

Figure 7.7
Simple pendulum

equilibrium position. The acceleration at any time is proportional to the distance from the equilibrium position.

In walking, every swing of the moving foot past the stationary foot is equivalent to one-half cycle of simple harmonic motion because the foot starts from rest, is accelerated forward at a rate roughly proportional to its distance from the stationary foot, passes the stationary foot at high speed and comes to rest as it touches the ground. The maximum velocity and acceleration during simple harmonic motion are given by

$$v_{max} = \frac{2\pi x_o}{T} \quad \begin{matrix} \text{(occurs at the} \\ \text{equilibrium position)} \end{matrix} \quad (7\text{-}15)$$

$$a_{max} = \frac{4\pi^2 x_o}{T^2} \quad \begin{matrix} \text{(occurs at the end} \\ \text{points of the swing)} \end{matrix} \quad (7\text{-}16)$$

EXAMPLE

A man walking takes 120 steps per minute each step being 1 metre in length. What is (i) the maximum velocity and (ii) the acceleration of the foot?

(i) The foot is at rest for 0.5 s, then advances 2 metres forward and comes to rest in the next half second.

$$v_{max} = \frac{2\pi(1)}{1}$$

$$= 6.2 \text{ m s}^{-1}$$

Note how large this velocity is compared to the advance velocity of the body as a whole which is

$$v = \frac{120}{60}$$

$$= 2 \text{ m s}^{-1}$$

(ii)

$$a_{max} = \frac{4\pi^2 1}{1}$$

$$= 39.5 \text{ m s}^{-2}$$

Note that this acceleration is approximately four times the acceleration due to gravity.

STANDING VERTICAL JUMP

To execute a standing vertical jump the jumper first crouches low and then pressing against the floor with his feet pushes upwards. Thus the jump consists of two parts: the first is termed the stretching segment (s) and is the part from the low crouch to the time when the feet leave the ground, the second is the free flight segment (h) when the body is completely airborne. The centre of mass of the body travels a total distance ($s + h$) during the jump. The average upward force on the jumper during the stretching segment is

$$F = F_R - mg$$

where F_R is the reaction force of the floor against the soles of the jumper's feet and mg is the gravitational force acting down. If the convention that up is positive is used, mg must have a negative sign. According to Newton's second law the acceleration of the jumper is

$$a = \frac{F}{m}$$

$$= \frac{F_R - mg}{m}$$

The velocity attained at the instant of lift off can be obtained from Equation (7-9) as

$$v^2 = 0^2 + \frac{2Fs}{m}$$

$$= \frac{2(F_R - mg)s}{m}$$

From Equation (7-9) the maximum height h is

$$h = \frac{v^2}{2g}$$

$$= (F_R - mg)\frac{s}{mg}$$

Note that the acceleration during free flight is just g acting down, air resistance being neglected. Greater height can be obtained by (i) crouching lower which increases the stretching segment, (ii) increasing the average reaction force F_R of the ground, (iii) executing the jump on a mountain top where the acceleration due to gravity is less or (iv) increasing the length of the legs which also increases the stretching segment s.

Many activities can be analyzed using the techniques just discussed. These analyses can lead to an understanding and improvement of the performance of the human body.

7-7 Work and Energy

Animals execute motion via muscular movement at the expense of chemical energy derived from the food they eat. By using up this energy they are able to perform various feats or in other words they do work. Work is defined in a physical sense as the product of a force and the displacement through which the force acts. If the displacement occurs at an angle to the line of action of the force only the component of the force along the displacement contributes, Figure 7-8.

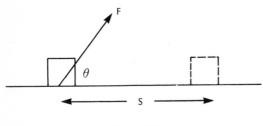

Figure 7.8

The definition of work

In general then

$$W = Fs \cos \theta$$

or

$$W = Fs \quad \text{if} \quad \theta = 0° \qquad (7\text{-}17)$$

Energy and work are interconvertible. If a body does work it does so at the expense of energy. If work is done on a body, energy can be stored within the system. There are many types of energy, chemical, electrical, etc., but we will be concerned here only with the two kinds of mechanical energy, kinetic energy and potential energy. Kinetic energy is the energy that a body possesses as a result of its motion and is defined as

$$E_K = \tfrac{1}{2}mv^2 \qquad (7\text{-}18a)$$

where m is the mass of the object and v is its velocity. For translational motion v is simply the linear velocity of the object. If in addition a rigid body is executing rotational motion about a fixed axis then there is a rotational kinetic energy contribution which is the sum of the kinetic energies of each individual particle given by

$$E_K = \sum_i \tfrac{1}{2}m_i v_i^2$$

using Equation (7-13)

$$v_i^2 = (\omega r_i)^2$$

therefore

$$E_K = \tfrac{1}{2}\sum_i m_i r_i^2 \omega^2$$

$$= \tfrac{1}{2}\sum_i (m_i r_i^2)\omega^2$$

from Equation (7-14)

$$I = \sum_i m_i r_i^2$$

thus

$$E_K = \tfrac{1}{2}I\omega^2 \qquad (7\text{-}18b)$$

In general then the total kinetic energy is a sum of Equations (7-18a) and (7-18b). Potential energy is the energy a body possesses as a result of some work that has been done on it.

In mechanics, one of the most important types of potential energy is gravitational potential energy which is the energy a body possesses due to its position and is given by

$$E_P = mgh \qquad (7\text{-}19)$$

where g is the acceleration due to gravity and h is the height of the body above the level taken as the zero of the potential energy. In most physical systems, energy is continually converted back and forth between the two forms. Of course in real systems much energy can be lost in the form of heat as the result of friction.

EXAMPLE

Using energy considerations, determine the free flight height reached in a standing vertical jump.

The work done on the jumper during the crouch segment of the jump is given by

$$W = \text{force} \times \text{distance over which}$$
$$\text{the force acts}$$
$$= (F_R - mg)\, s$$

As the jumper is accelerated upwards this work is converted into kinetic energy. At the maximum jump height, the jumper momentarily comes to rest and the total energy is now in the form of gravitational potential energy given by

$$E_P = mgh$$

$$\text{Work done} = \text{energy acquired}$$

Therefore

$$mgh = (F_R - mg)\, s$$

and

$$h = \frac{(F_R - mg)}{mg}\, s$$

7-8 Stress and Strain

Often in mechanics, the term rigid body is used and up until now we have considered all objects as rigid. In actual fact there is no such thing as a rigid body. All material deforms to some extent under the action of applied forces. The ability of a material to withstand deformation without breaking or becoming permanently deformed is termed its elasticity. Bone, muscle, tendon—the structural elements of the human body—vary greatly in their ability to withstand applied forces without fracture or tear. Recently, there has been a growing interest in the elastic properties of these biomaterials.

Suppose that a uniform rod of a biomaterial, perhaps collagen, of length ℓ and cross-sectional area A is subjected to equal but opposite forces $\vec{F}$ applied to its ends, Figure 7-9.

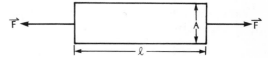

Figure 7.9
Rod of biomaterial under tension

The rod is said to be under tension (or compression if the forces are directed inwards). The tensile stress (σ) on the rod is defined as

$$\sigma = F/(A - \Delta A) \qquad (7\text{-}20)$$

where ΔA is the change in the cross-sectional area. For some materials ΔA is extremely small and can be neglected with respect to A. For biomaterials ΔA can be significant. It can easily be demonstrated from Equation (7-20) that σ has the units of pressure. Under the action of this tensile stress the rod will deform, that is its shape and dimensions will change. Strain (ε) is defined as the relative change in the dimensions of the rod, Figure 7-10.

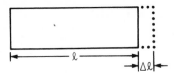

Figure 7.10

Deformation of rod under tension

Thus if ℓ is the original length of the rod and $\Delta\ell$ is the elongation caused by the applied stress,

$$\varepsilon = \frac{\Delta\ell}{\ell} \qquad (7\text{-}21)$$

which of course is dimensionless being simply the ratio of a length to a length. In rods or fibres of biomaterials, the molecular processes occurring during stretching may be quite complicated involving uncoiling of chains, breaking of cross-linked bonds, etc.

Within the elastic limit of the material the strain is proportional to the stress.

$$\sigma \propto \varepsilon \qquad (7\text{-}22)$$

A typical stress-strain graph is shown in Figure 7-11. For the elastic region, the relationship is linear and

$$\sigma = Y\varepsilon \qquad (7\text{-}23)$$

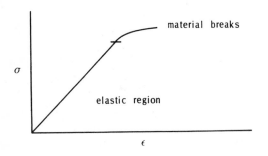

Figure 7.11

Typical stress – strain relationship

where Y is known as the Young's modulus of the material. Y has the dimensions of stress since strain is dimensionless.

Equation (7-23) is often called Hooke's Law. The Young's modulus which can be thought of as a measure of the stiffness of a material is known for many materials, Table 7-2. Note the great range of values of Young's modulus for biomaterials. A complication often arises with many biomaterials in that the strain is time dependent which means that the fibre stretches slowly for a long time after the stress is applied. The strain can also show a hysteresis effect.

Table 7-2
Young's Modulus for Various Materials Under Compression

Material	$Y(N\ m^{-2})$
Bone (along axis)	2×10^{10}
Tendon	2×10^{7}
Rib cartilage	1.2×10^{7}
Blood vessels	2×10^{5}
Steel	2×10^{11}

EXAMPLE

A strip of tissue 5 cm long with a cross-sectional area A of $0.10\ cm^{2}$ is cut from the wall of the aorta. Such material has a Young's modulus of approximately $2 \times 10^{5}\ N\ m^{-2}$. What mass suspended from the strip hung vertically will cause a 0.5 cm elongation if ΔA can be assumed to be negligible? From Equation (7-20) (neglecting ΔA)

$$\sigma = \frac{F}{A}$$

$$= \frac{mg}{10^{-5}}\ N\ m^{-2}$$

where mg is the weight of the suspended mass. From Equation (7-21)

$$\varepsilon = \frac{\Delta \ell}{\ell}$$

$$= \frac{0.5}{5}$$

$$= 0.1$$

From Equation (7-23)

$$\sigma = Y\varepsilon$$

thus

$$\frac{mg}{10^{-5}} = (2 \times 10^5)(0.1)$$

$$m = \frac{(2 \times 10^5)(0.1)(10^{-5})}{10}$$

$$= 2 \times 10^{-2} \text{ kg}$$

Many structural components of biological systems are assembled from a variety of substances rather than just one, such as in a wire. This can give rise to some rather unusual stress-strain diagrams. Consider the ligamentum nuchae which is a very heavy "cable" along the top of the neck of cows and other ungulates. It must be strong to hold a very heavy head when grazing. It should yield readily at first if subjected to a sudden stress but should "tighten-up" before the strain becomes excessive. It must be a biological shock absorber. The actual stress-strain diagram shows exactly these desirable properties (Figure 7-12, solid line).

The two major structural components are the proteins elastin and collagen. If the collagen is removed in vitro, by the action of the enzyme collagenase, the resultant stress-strain relationship indicates that the elastin is principally responsible for the initial high yield response of the intact ligament. Removal, on the other hand, of elastin by elastase leaves collagen which is seen to be responsible for the

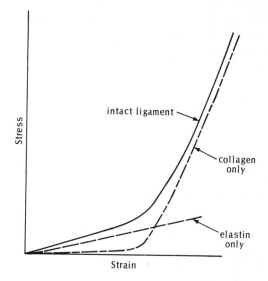

Figure 7.12

Stress-strain diagram for the ligamentum nuchae

final portion of the response of the ligament. Obviously the collagen has a greater Young's modulus than the elastin. Why if the collagen is relatively inextensible, does it have no effect when stress is first applied? Structural analysis at the microscopic level indicates that the collagen fibres are slack at zero stress and can have no effect until the elastin yields enough to take up the slack from the collagen.

Another type of deformation arises when a force is applied tangentially to a surface, for example when a bone is twisted in skiing resulting in a spiral fracture of the tibia. Now

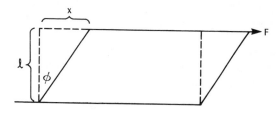

Figure 7.13

Deformation under shear stress

instead of the atoms in the material being just pulled apart, layers of atoms are forced to slide one over the other. This type of stress is termed a shear stress. First consider the simplest case (Figure 7-13) where a force F acts over the entire top surface A of a block of material which is anchored at the bottom. A shear stress

$$\sigma_s = \frac{F}{A} \qquad (7\text{-}24)$$

causes a shear strain

$$\varepsilon_s = \frac{x}{\ell} = \tan \phi = \phi \qquad (7\text{-}25)$$

(since ϕ is a very small angle).

The ratio of the shearing stress to the shearing strain is called the shear modulus G.

$$G = \frac{\sigma_s}{\varepsilon_s} \qquad (7\text{-}26)$$

Table 7-3 lists values of the shear modulus for various materials which exhibit an elastic region with respect to shear.

Table 7-3
Shear Moduli for Various Materials

Material	$G(\text{N m}^{-2})$
Long Bone	$0.8\text{--}1.5 \times 10^{10}$
Steel	8×10^{10}
Aluminium	2.6×10^{10}
Rubber (soft, vulcanized)	1.6×10^{6}

Now consider what happens when shear stresses are applied to one end of a solid rod (Figure 7-14(a)) which is fixed at the other end. For the solid cylinder:

$$\frac{F}{A} = G \left(\frac{x}{\ell}\right)_{\text{avg}}$$

at the centre $x = 0$

$$\therefore \left(\frac{x}{\ell}\right)_{\text{avg}} = \frac{x}{2\ell}$$

The area A is that of the circular end,

$$\therefore A = \pi r^2$$

$$\therefore F = \frac{\pi r^2 G x}{2\ell} = \frac{\pi r^3 G x/r}{2\ell}$$

but $x/r = \theta$ radians and since $Fr = M$ the moment or torque exerted on the end

$$\therefore M = \frac{\pi G r^4 \theta}{2\ell} \qquad (7\text{-}27)$$

For a hollow cylinder, (Figure 7-14(b)) which more nearly approximates a long bone, we assume that the thickness t is small and that there is an average radius

$$\bar{r} = \frac{r_i + r_o}{2} \approx r_o$$

The area here will be $2\pi \bar{r} t$ and, since t is small, all particles at the end rotate the same distance x. Using a similar analysis it can then be shown that for the hollow rod

$$M = \frac{2 G \pi \bar{r}^3 t \theta}{\ell} \qquad (7\text{-}28)$$

The material down the centre of a cylinder is deformed little and therefore contributes little resistance to twist. The hollow bone is therefore almost as strong as a solid one and has the advantage of being much lighter.

7-9 Scaling

Giant creatures and miniature people are subjects of many a horror movie or children's story. Could such scaled up or down creatures really exist? The interesting and instructive approach to animal mechanics espoused by

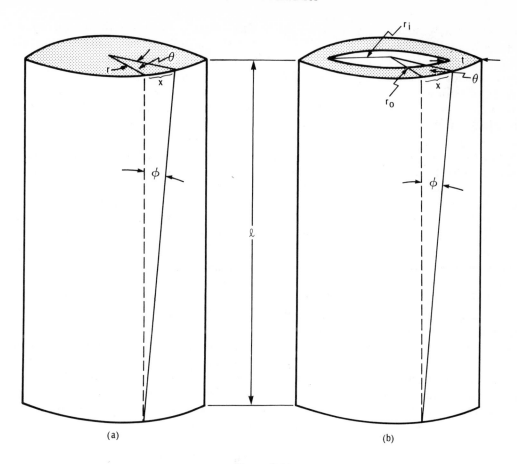

(a)   (b)

Figure 7.14
Shear deformation of solid and hollow cylinders

scientists like D'Arcy Thompson permits the avoidance of a rigorous quantitative argument while presenting a qualitative approach that is usually quite adequate. This is the process of scaling, which is a type of dimensional analysis. All that is.really necessary is a knowledge of the types of forces acting.

It has already been seen that mammals like man live in an environment where the prime force is gravity. Insects, on the other hand, may be constrained by a force of which man is hardly aware, surface tension (see Chapter 9).

A man emerging from a bath carries a film of about a half kilogram of water but a wet bug carries many times its own weight. Consider how constrained your movements would be if you carried many times your own weight of water. Bacteria exist in a world where viscosity (Chapter 9) and Brownian motion (Chapter 10) of the fluid environement are important. Several examples will be considered to demonstrate the power of scaling arguments. A review of dimensions, as presented in Appendix 2, would prove useful to many readers.

Have you ever wondered why there are no land creatures larger than the elephant? The answer is that the size of a free standing mammal cannot exceed certain limits using the material nature has provided. There are two restricting limits: static limits and dynamic limits. First consider the static limits. The resistance of the supporting parts of the body, the limbs, to a crushing stress is proportional to their cross-sectional area. Thus if a small animal with leg radius r_1 were scaled up in all dimensions by a scaling factor L, such that the new leg radius were Lr_1 then the area would have increased by a factor L^2 and the strength would have increased by L^2. It can be shown that leg strength is proportional to L^2 whether the leg is pillar-like as in the elephant or "folded" as in a kangaroo. The legs however have to hold up a body the volume and weight of which will increase by L^3. If an animal were scaled up by $L = 10$ then legs 100 times as strong as before would have to hold up 1000 times the weight and collapse would be inevitable. As animal bodies get larger the legs must get relatively thicker with the diameter increasing as $L^{3/2}$. Thus the small deer can have very spindly legs whereas much of the space under the elephant is occupied by legs. However, if the body weight is balanced by a counter force, for example the buoyant force of water, mammals can attain a greater size than the strength of body materials would normally permit. Thus the enormous size of some whales can be accounted for.

There are also dynamic considerations that limit the size of an animal. An animal must move if it is to survive. To move a limb, an animal must do work. The work required to move a limb of mass m a distance s is proportional to ms. Now $m \propto L^3$ and in normal situations a limb can only be moved a distance of the order of its own size, therefore s is of the order of L and consequently the work required will be $W \propto ms \propto L^4$. To provide this work a muscle must exert a force F through a distance s. But the force exerted by a muscle will be proportional to L^2 (if this bothers you compare a weight-lifters muscles with your own—they are both the same length but his are greater in cross-sectional area) and s again will be of the order of L. Therefore the work that can be done will increase as $W \propto Fs \propto L^3$. Here too, as the animal is scaled up, it gets into trouble since the work required to move the limb increases faster than the ability of the animal to provide the work.

Earlier in the chapter the standing vertical jump was investigated. It can be shown, using scaling arguments that a small and large animal of like form ought to be able to jump to approximately the same height. As before the maximum work the muscles can do will be proportional to L^3. This work must lift a mass (proportional to L^3) a distance h. The work required then is proportional to L^3h and consequently h must be essentially constant for all similarly constructed animals, regardless of their size. A grasshopper and a flea can jump to about the same absolute height but the grasshopper jumps to 20 times his own height while the flea jumps to 200 times his.

With this information you can console yourself with the thought that you are about as good a vertical jumper as the grasshopper. Moreover you really wouldn't want to jump several times your height or your legs would be crushed under your own weight when you came down.

Using similar arguments it can be shown that similarly constructed animals of different sizes will have about the same top speed on level ground but that the larger one will always lose out when going uphill. Thus a horse can match a dog on the level but quickly drops behind when a climb is encountered.

These are but a few of the types of problems that can be approached using the tools of dimensional analysis. The same types of analyses can be used if an animal is scaled down, i.e. if L were less than one. Think about what would happen to the frequency response of your ear or the far point of your eye if all your body dimensions were scaled down by $L = 0.1$.

Problems

1. What are the horizontal and vertical components of the 20.0 N force shown in the diagram?

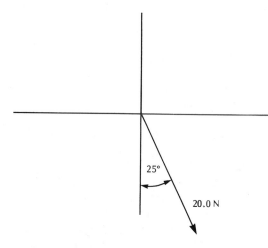

2. A butterfly is flying in a garden. First it travels due south for 6.0 m and then 37° W of N for 10 m. Assuming that the positive x-direction is east, and the positive y-direction is north,
   (a) what are the coordinates of the butterfly's final position relative to its initial position?
   (b) what are the magnitude and direction of the displacement of the butterfly (relative to its initial position)?

3. A force acts at 30° to the horizontal. Its horizontal component is known to be 200 N. What are its magnitude and its vertical component?

4. A bird is flying due north with a speed of 15 m s^{-1}. A train is travelling due east with a speed of 15 m s^{-1}. What are the magnitude and direction of the velocity of the bird from the point of view of a passenger in the train?

5. A bird is migrating non-stop a distance of 2000 km. Its destination is due south of its present position. In still air, the bird can fly with a speed of 40 km hr^{-1}. However, there is a wind of 25 km hr^{-1} from the west.
   (a) In order to make the trip following a straight line path, in what direction should the bird head?
   (b) How long will the trip take?

6. A boat is going to head due north but unfortunately the tide is running out (east) with a speed of 10 m s^{-1}. If the boat can travel with a maximum speed of 15 m s^{-1} in still water, what is the maximum speed with which it can head due north relative to the shore?

7. The force F acts $\frac{3}{4}$ of the way along the boom of length 20 m. What is the torque (moment) of this force about the point O?

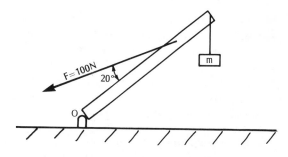

8. A teeter-totter is 20 kg in mass. Assume it is a uniform plank of length 2 m. The pivot point is placed 0.2 m from the midpoint. What weight must be added to the short end to balance the teeter-totter?

(Hint: Take moments about the pivot—the uniform plank can be assumed to be a single weight concentrated at its midpoint. Can you justify this?)

9. A man is standing as shown. His upper body weighs 1000 N (i.e. 100 kg or 220 lbs)! and has a centre of gravity C about 0.6 m from an effective pivot point O. Suppose the muscles in his back which maintain the position are a perpendicular distance of 6 cm from the pivot point. What force are they exerting?

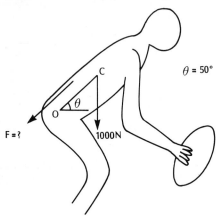

10. For the example on the tension in the deltoid muscle of the arm determine how the expression for the tension is changed if a weight W_1 is placed in the hand.

11. A 1.0 kg crow sits on a telephone wire midway between two poles 50 m apart. The wire, assumed weightless, sags by 0.1 m. What is the tension in the wire?

12. A 100 kg man stands 1 m from the end of a 4 m scaffold which weighs 750 N. What are the tensions T_1 and T_2 in the supporting ropes?

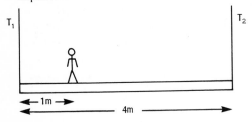

13. By modifying the example on the standing vertical jump determine an expression for the vertical height attained during a jump at an angle θ to the horizontal.

14. What force F_m would the muscle exert to support the 10 kg mass? Neglect the mass of the forearm. What force does the bone in the upper arm (humerus) exert on the forearm?

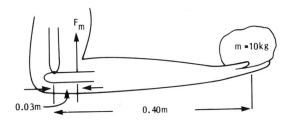

15. Suppose the arm in question 14 is thrust forward until the upper arm makes an angle of 30° from the vertical. The forearm stays level. Determine the new values for the quantities sought in question 12.

16. A cat leaps horizontally from a tree branch 3 m above the ground. It lands at a point 4 m horizontal from the point below which it leaped. What is the vertical component of the cat's velocity at the instant it touches the ground?

17. What horizontal force will stop, in 4 seconds, a 5 kg mass sliding on a frictionless floor if the mass has an initial speed of 12 m s^{-1}?

18. The great swimmer, Mark McSpitswater, dives from a diving board 15.0 m above the water's surface. His initial velocity is 2.00 m s^{-1} at an angle of 30.0° up from the horizontal.
    (a) How long does it take him to hit the water? (recall that $x = (-b \pm \sqrt{b^2 - 4ac})/2a$ is the solution to $ax^2 + bx + c = 0$.)
    (b) As he hits the water, what is the magnitude of his displacement from the diving board?

19. The mass in the diagram slides along a horizontal frictionless floor, pulled by the force F as shown. What is the acceleration?

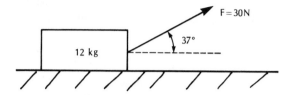

20. A 25 kg mass is acted on by a constant 5 newton force. If the object started from rest, how far would it have travelled after 4 seconds?

21. A horse pulls an old-fashioned cutter along a level icy surface at a steady speed of 15 m s^{-1}. The cutter and its load have a total mass of 250 kg. If the horse is providing a constant pulling force of 125 newtons, what is the coefficient of friction?

22. A high jumper jumps 1.2 m straight up. With what speed did he leave the ground?

23. A 10 N force at an angle of 35° to the horizontal is applied to the 5 kg block. In addition there is a 7 N frictional force acting as shown.

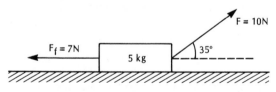

(a) Find the work done by each of the *four* forces acting on the block in moving it 12 m horizontally.
(b) What is the total work done?
(c) If the block starts from rest, what is its speed after moving the 12 m?

24. A baseball player running with speed $v = 9 \text{ m s}^{-1}$ slides into second base through a distance of $d = 4$ m. Find μ, the coefficient of friction.

25. In a normal joint the coefficient of friction is approximately 0.003. In arthritic patients the value of this coefficient can be considerably larger because of changes in the lubricating fluid. Assuming all other things equal how much more energy is dissipated in an arthritic joint where $\mu = 0.03$ than in a normal joint?

26. You are walking along a wood floor. During the walking process your heel strikes the floor with a force $\vec{F}$ at an angle θ to the vertical. If $\mu_s = 0.5$ for normal shoe leather on wood what is the maximum value that θ can take and still ensure that you do not slip?

27. A donkey is used to raise water from a well. He is walking around a circle of 2 m radius at a speed of 3 km hr^{-1}. He is raising 100 kg of water a distance of 2 m every second. What force is he exerting on the lever arm?

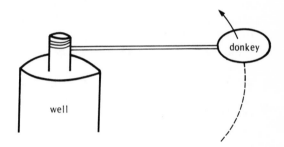

28. A nurse wants to inject 2 g of polio vaccine into a student. With what average force must she push on the plunger if the speed of the vaccine coming out is 1 m s^{-1} and the length of the plunger's motion in the syringe is 6 cm (friction is ignored).

29. A lab technician is centrifuging a blood sample at an angular speed of 3000 rpm (revolutions per minute). If the radius of the circular path followed by the sample is 0.15 m, find the speed of the sample in m s^{-1}.

30. Assume a man's arm is 0.6 m long and that he can rotate it about his shoulder in a vertical circle once per second. Com-

pare the centripetal force developed with the gravitational force on the blood in his hand. (Note that the two forces will be in the same ratio as the accelerations.)

31. In a high dive, why does the human body in a tucked position rotate so much faster than in an extended position?

32. A figure skater is rotating at 0.64 revolution per second.
    (a) What is her angular velocity in radians per second?
    (b) If she now pulls her arms in so that her moment of inertia is reduced by $\frac{1}{3}$, what is her new angular velocity?

33. A uniform, solid sphere of mass m and radius r rolls down a hill of height 20 m. Ignoring friction what is the sphere's speed at the bottom? (I of sphere = $\frac{2}{5} mr^2$).

34. For the human tibia bone the ultimate tensile strength (the point at which fracture occurs) is $1.4 \times 10^8 \, \mathrm{N\,m^{-2}}$. What is the strain at the point of fracture? Assume that Hooke's Law is valid up to the point of fracture.

35. When standing erect, a person's weight is supported chiefly by the larger of the two leg bones. Assuming this bone to be a hollow circular tube of 2.5 cm internal diameter and 3.5 cm external diameter, what compressive load must this bone (in each leg) carry in the case of a 70 kg person?

36. If the leg bones described in question 35 are 90 cm long what is the amount of compression in each leg at the fracture point? Assume that the leg bones remain elastic under compression until fracture, which occurs at a stress of $1.4 \times 10^8 \, \mathrm{N\,m^{-2}}$.

37. If the person in question 35 is in a skiing accident where one end of the bone described is twisted 10° further than the other, what torque must have been ap-

plied? Assume the shaft of the bone is 0.30 m long.

38. A sample of bone in the form of a cylinder of cross-sectional area 1.5 cm^2 is loaded on its upper end by a mass of 10 kg. Careful measurement with a travelling microscope reveals that the length decreases by 0.0065 percent. What is Young's modulus for the specimen?

39. Suppose you are supported by a single steel wire. (Ouch!) What must be the diameter of the wire if the elongation is not to exceed 0.1 percent?

40. How does the maximum work done by animals of similar shape scale in terms of size L?

41. A cell's oxygen requirement is proportional to its mass, but its oxygen intake is proportional to its surface area. Show that a cell cannot grow indefinitely and survive.

42. If it takes 100 g of wool to knit a sleeveless sweater for a 15 kg child, approximately how much wool is required for a similar sweater for a 120 kg man? Assume that all dimensions on the man are a simple multiple of those on the child.

43. If a giant species of ant were developed with all body measurements (excluding legs) 10 times as large as the common ant then what would the diameter of the legs, in order to be relatively as strong as in the common ant, have to be?

44. A 70 kg man and a 0.5 kg hairless dog go swimming. When they emerge 0.5 kg of water adheres to the man's skin. How much adheres to the dog's? What percentage of the mass of each is water?

45. In a circus act a pony and a horse are balanced on a teeter-totter. If the pony is 12 hands high and the horse is 18, then what will be the ratio of the distances $\dfrac{\text{Pony to pivot point}}{\text{Horse to pivot point}}$ (approximately)?

Chapter 8
Molecular Biophysics

8-1 Introduction

In some ways biological systems appear fairly simple. For example, only about one-half dozen types of biological macromolecules are required for the structure and operation of the cell. However, the variability in any one type of biological macromolecule is enormous. Exactly how this variability arises from chemically similar molecules has been one of the most puzzling scientific problems of recent years. Since proteins provide the best example of variability, they will be considered more deeply in the following pages. The principles which are discussed are generally applicable to other macromolecular types as well.

8-2 Molecular Interactions and the Role of Water

Some of the basic types of interactions which can influence the conformations of individual molecules or the interactions between molecules will first be considered. They will then be illustrated with specific examples, principally as they apply to proteins.

Electrostatic interactions may be attractive (between unlike charges) or repulsive (between like charges). Sometimes referred to as Coulomb forces, they are one of the most important types of interactions. Charges at different points in the same molecule may influence its configuration or permit it to interact with solvent molecules, other molecules

in solution or with ions in solution. Relative to the other forces to be considered, electrostatic interactions act across considerable distances and are referred to as long-range forces. If r is the distance between charges of q_1 and q_2 coulombs, then the force between the charges will be:

$$F = \frac{kq_1q_2}{Kr^2} \qquad (8\text{-}1)$$

where $k = 9 \times 10^9 \, \text{N m}^2 \, \text{C}^{-2}$ and K, a dimensionless number, is characteristic of the medium between q_1 and q_2 and is called the dielectric constant. For free space or air, K is approximately 1 but for water it is about 80. A table (13-2) of other dielectric constants is provided later in the book. One coulomb is really a very large amount of charge. More frequently the charges will be integral values of the charge on the electron, $1.6 \times 10^{-19} \, \text{C}$. Note that when charged molecules are dissolved in water the effect of the medium is to make the force between them only 1/80 of what it would have been in free space. This helps in dissolving many crystalline materials such as salt. If, however, the molecules do come very close together so that water molecules are squeezed out from between them, then the force can suddenly increase rapidly as the medium then approximates free space.

As will be shown in detail later (Chapter 13) the potential energy of the interaction is:

$$W = \frac{kq_1q_2}{Kr} \qquad (8\text{-}2)$$

When the charges are opposite, ions or molecules are attracted towards each other but only until the electron clouds start to overlap. Once this occurs, a repulsion quickly develops which increases very rapidly if r decreases still more. The attracted groups will tend to vibrate about an average separation corresponding to the minimum potential energy of the system.

EXAMPLE

At some instant, two molecules in water are separated by 3.5 nm. One has an excess electron and the other lacks an electron. What is the energy of the interaction? Equation (8-2) can be used to obtain:

$$W = \frac{9 \times 10^9 \, \text{N m}^2 \, \text{C}^{-2}(+1.6 \times 10^{-19} \, \text{C})(-1.6 \times 10^{-19} \, \text{C})}{80 \times 3.5 \times 10^{-9} \, \text{m}}$$

$$= -8.2 \times 10^{-22} \, \text{N m} = -8.2 \times 10^{-22} \, \text{J}$$

The negative sign indicates that the interaction is attractive. Note that as r becomes very large W approaches zero.

Frequently interactions are given on a molar basis. To do this the answer above would be multiplied by Avogadro's number.

Dipoles occur whenever a charge separation occurs within a molecule or portion of a molecule that has no net charge. For example the carboxyl group (Figure 8-1(a)) has a slightly negatively charged oxygen ($\delta-$) and a slightly postively charged carbon ($\delta+$). The dipole formation in this case is due to the fact that the electron affinity of oxygen is greater than that of carbon. The essential properties of the dipole charge distribution are the magnitude of charge q and the separation d between the charges. The product qd is called the dipole moment. Molecules or groups

within molecules which have a dipole moment are called polar. The interaction energy between a pair of isolated permanent dipoles decreases as $1/r^3$. The situation becomes more complicated for permanent dipoles in solution where they are subjected to random thermal bombardment by other molecules. In solution the interaction energy goes as $1/r^6$ for weak dipoles or for strong dipoles at large separations. These forces have, therefore, much shorter ranges than electrostatic forces. If two dipoles are placed near each other they will tend to twist about so that opposite charges come as close together as possible (Figure 8-1(b)).

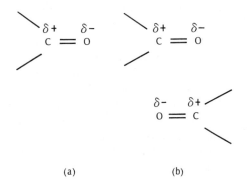

(a)   (b)

Figure 8.1

Dipolar interaction

If a permanent dipole and a non-polar molecule approach each other, the charges in the permanent dipole may induce, due to attraction and repulsion, a redistribution of charge within the non-polar molecule with the result that it becomes a transient dipole. The two molecules can then attract each other. The energy of interaction varies as $1/r^6$ in this instance also.

Two molecules, both of which are non-polar, can even attract each other. Random motion of charge within one molecule may, at some instant, cause it to be a transient dipole.

The transient dipole can interact with a neighbour to induce a dipole moment. A transient dipole-induced dipole interaction can then occur and, like the other types of dipole-dipole interaction, this has an energy that decreases as $1/r^6$. At short distances, the strength of this last interaction can be greater than the interaction between permanent dipoles. The fluctuations of the transient dipoles occur at frequencies corresponding to those of light. The force is often referred to as a dispersion force or more specifically, since it was proposed by London, the London dispersive force. Van der Waals interactions are also dipole–dipole interactions.

An extremely important but short-range interaction is the so-called hydrogen bond. This "bond" is partly ionic and partly covalent (sharing of electrons between atoms) in nature and has an energy of about 10 percent of an average covalent bond, or about $1-4 \times 10^4 \, J \, mol^{-1}$. Because the energy of a single hydrogen bond is comparable to the available thermal energy, in solution hydrogen bonds can easily be made or broken at physiological temperatures. However, since some molecules contain hundreds of hydrogen bonds, a significant amount of energy may be involved in this interaction and the resulting structure can be quite stable. Each hydrogen bond consists of a donor group and an acceptor group as shown in Figure 8-2. Hydrogen bonds are highly directional. By this it is meant that the sequence of atoms C, O, H, and N as in Figure 8-2 is usually, but not always, a straight line.

Acceptor Group   Donor Group

Figure 8.2
Hydrogen bond

The final type of interaction to be considered is the molecule-solvent interaction. Since the solvent of interest will almost always be water its "structure" should be considered before this final interaction can be fully appreciated.

It is well known that crystalline solids possess a characteristic structure or "arrangement of molecules." The molecules in such a solid are arranged in a regular three-dimensional array or lattice structure. Of course they are not perfectly static—the molecules vibrate about their mean lattice positions and occasionally gain enough energy to diffuse to new positions.

Liquids also possess some degree of structure. The molecules of a liquid are essentially as close together as in a solid and so still interact by electrical forces. Thus one molecule can influence the position and orientation of its nearest neighbours. In the liquid, the molecules possess more thermal energy than in a solid and hence "break away" from these interactions with their neighbours and diffuse to new positions much more frequently than do the molecules of a solid. Because at any instant, many molecules are undergoing this diffusion process the structure of a liquid lacks the long-range order. That is, order over a distance of many molecular diameters may exist in solids but not in liquids.

Because of its inherent interest and possible importance in problems related to the solubility of materials in aqueous solutions (e.g. hydrophobic and hydrophilic bonds), the structure of liquid water has been intensively studied. Many models and proposals for the structure have been presented. However none of these models can explain all of the known properties of water. Much controversy exists and the matter is still undergoing intensive research. Some of these ideas are now presented.

To understand the current ideas about liquid water, an isolated water molecule must first be examined. Its shape is fairly well known from spectroscopic studies of water vapour and from quantum mechanical calculations. The isolated molecule is essentially V-shaped. The 8 electrons of the oxygen atom are arranged in a ls orbital (2 electrons) and in 4 sp^3 hybrid orbitals (6 electrons). The latter are similar to the well known sp^3 orbitals of carbon and are arranged in a symmetric tetrahedral pattern about the oxygen nucleus.

In oxygen, the 6 outer electrons are placed in these 4 sp^3 orbitals, 2 electrons in 2 of the orbitals and 1 electron in each of the other two orbitals. Each of the latter two orbitals could accept one more electron. Thus when an H_2O molecule forms, the ls electron from one hydrogen overlaps one of the unfilled sp^3 orbitals of oxygen to bind the hydrogen covalently to the oxygen. Similarly, the second hydrogen bonds via the remaining unfilled sp^3 oxygen orbital. The result is shown in Figure 8-3. The angle between the OH bonds is found to be 105° and the OH distance is 0.099 nm.

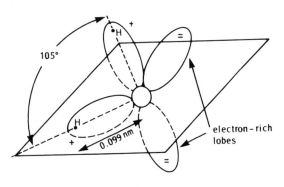

Figure 8.3
Tetrahedral structure of a water molecule

Because of the 8 positive charges on the oxygen nucleus compared to only 1 on the hydrogen nucleus, the electrons in the OH covalent bond tend to be drawn toward the oxygen nucleus leaving the H nuclei or pro-

tons somewhat "bare" of electrons and hence possessing a "positive bias." The oxygen end of the molecule tends to have a "negative bias" directed along the two originally filled oxygen sp^3 orbitals. These electron-rich "lobes" are shown in Figure 8-3, on the side of the molecule remote from the hydrogens and directed above and below the plane of the molecule. The molecule thus possesses a dipole moment or is polar, the hydrogen nuclei being the positive end and the oxygen the negative end of the dipole.

When H_2O atoms are brought close together without too much kinetic energy (as in ice formation) there are fairly strong electrical interactions between the positively biased H nuclei of one water molecule and the negatively biased electron-rich lobes of oxygen in adjacent water molecules. Thus a bond called a "hydrogen bond" forms between the hydrogen of one water molecule and the oxygen of an adjacent molecule. Although considerably weaker than regular covalent bonds, the hydrogen bonds are sufficient to hold the molecules together to form ice. Each water molecule bonds to four others, two via its own two hydrogens and two via its two electron-rich lobes. Each water molecule is thus surrounded by four nearest neighbours in a tetrahedral arrangement as shown in Figure 8-4 which is the characteristic structure of ice. The distance between a H nucleus and its hydrogen bonded oxygen nucleus is 0.177 nm making the overall O-O distance in ice about 0.276 nm. This arrangement results in a very open cage-like structure for ice—there is a good deal of empty space within the ice lattice. Thus ice has a relatively low density when compared to many other solids and compared to liquid water.

When ice melts, energy must be added to it—about $3.35 \times 10^5 \, J \, kg^{-1}$. However this energy is sufficient to break only about 20 percent of the hydrogen bonds originally in

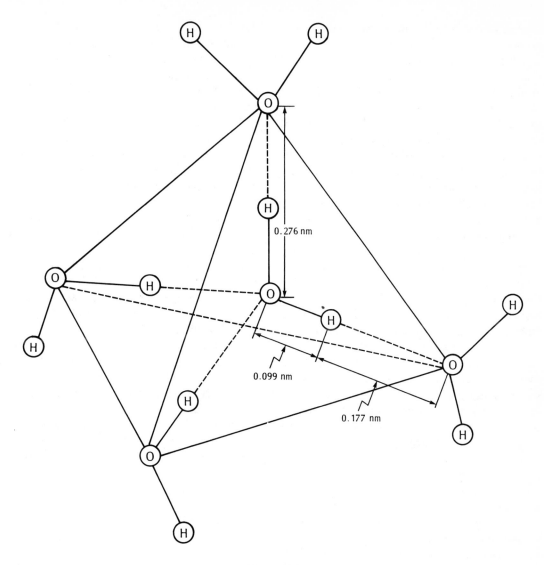

Figure 8.4
Tetrahedral arrangement of water molecules in ice. (from F. M. Snell, S. Shulman, R. P. Spencer and C. Moss. Biophysical Principles of Structure and Function, Addison-Wesley Publishing Co. Inc., 1965.)

the ice. Thus an important fact about liquid water is that at any instant a fairly large fraction of the water molecules are hydrogen bonded to each other. However, these bonds are weak enough (say 2.1 to 3.7×10^4 J mol^{-1}) and the thermal energy of the water is large enough that the bonds are continually being broken and reformed so that a given molecule is sometimes "free" and at other times hydrogen bonded to its neighbours.

All the models proposed for liquid water utilize the above idea that many of the molecules are hydrogen bonded. Some models, known as "uniformist models" or "distorted hydrogen bond models," visualize all the molecules as four-coordinated, hydrogen bonded as in ice but with the hydrogen bonds bent or distorted to varying degrees from the "straight-line" arrangement of ice (as in Figure 8-4) to bonds which are so bent that they are essentially broken bonds. This results in irregular networks of linked molecules in the liquid. Another group of models which have been widely accepted are the "mixture and interstitial" models. These models propose that liquid water consists of from two to five distinctly different molecular arrangements or components and that the liquid is a mixture of these components. For example, the simplest of these models proposes that the liquid has two components. One of the components consists of water molecules hydrogen bonded into an ice-like structure whereas the second component consists of unbonded or free water molecules packed together in a random and more dense arrangement than in ice. In this model the tiny ice-like regions would float about in the dense regions and the relative number of molecules in each component would change with temperature thus yielding a change in density with temperature. The more complicated mixture models have components of 1-, 2-, and 3-bonded molecules, as well as the 0- and 4-bonded ones of the single 2-component model. Some have non-bonded molecules bouncing about in the spaces of the ice-like regions—these are known as interstitial positions and hence the name interstitial model.

In summary, there have been numerous models proposed for the structure of liquid water. None, to date, can account for all of the known properties of water. A common feature of all the models is that a large fraction of the molecules in liquid water are at any instant hydrogen bonded to some extent (from 0 to 4) to their neighbours with the bonds being continually broken and reformed.

The solvent in biological systems is usually water. The enormous complexity of water may be realized by observing that it is an ideal molecule for both dipole–dipole interactions and hydrogen bonding. The large attractive forces water feels for itself can be interrupted only by a molecule which also has large electrostatic, dipolar, or hydrogen bonding capability. That is, for a molecule to get "wet" when placed in water it must effectively complete for water molecules. When a molecule has these properties, and is "wettable" it is said to be hydrophilic, a word of Greek origin meaning "liking water."

On the other hand, there are many molecules like benzene, C_6H_6 (Figure 8-5); methane, CH_4; ethane, C_2H_6, etc. which are uncharged, have very small dipole moments, and are unable to take part in hydrogen bonds. These molecules cannot compete for water molecules and are, therefore, insoluble and are termed hydrophobic, "disliking water."

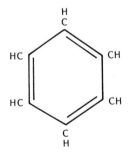

Figure 8.5
Benzene

Some molecules, as will be discussed in greater detail in the next section, have both hydrophilic and hydrophobic portions. Consequently when placed in water they will twist into a configuration in which as many hydrophilic groups as possible are on the outside, many hydrogen bonded to water molecules, and as many hydrophobic groups as possible are buried in the interior away from the water.

8-3 Protein and Nucleic · Acid Structure

Proteins constitute a very important group of biological macromolecules. There are many different proteins in a living cell. Each is specified by a different gene, and has a specific role to play in the normal function of the cell. The protein may be, for instance, an enzyme to regulate the rate of some important biochemical reaction, a carrier such as hemoglobin in the blood, a structural element, or a part of a contractile mechanism as in muscle. Different proteins have a wide range of molecular weights averaging around 40,000. Like many of the other important macromolecules they are polymers. They are assembled initially as linear heteropolymers. In a very few instances (e.g. the protein elastin) they may be subsequently modified into branched and cross-linked configurations. This contrasts with commercially made polymers which are often branched or cross-linked.

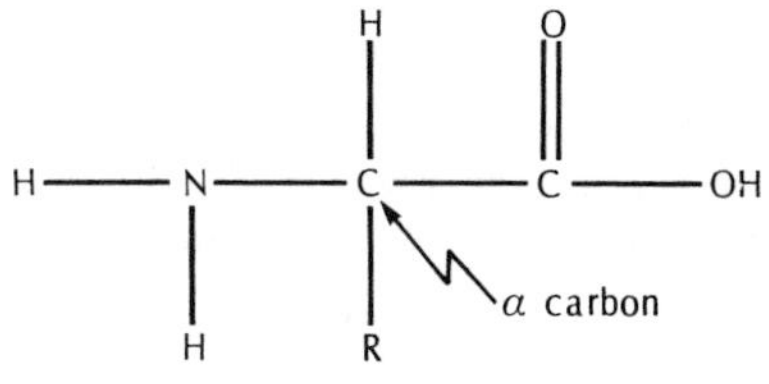

bon, to which are attached a carboxyl group ($-COOH$) an amino group ($-NH_2$), a hydrogen atom ($-H$) and a group called the R group which specifies the particular amino acid. This is shown in Figure 8-6.

In the initial assembly of proteins some 20 different kinds of amino acids, each with its own particular R group, are used. Some may be modified after the initial chain is assembled so that slightly more than 20 R groups appear in the final product.

As each amino acid is added to the chain a water molecule is removed and a peptide bond is formed. This is indicated in Figure 8-7 where the peptide bond is the carbon-nitrogen bond in the area enclosed by broken lines. The group enclosed in this area will be planar because the electron orbitals are distributed over the group in such a way as to shorten the peptide bond and give it partial double bond characteristics thus preventing rotation about this bond. Rotation about bonds in the chain can then only occur on either side of the alpha carbon and this helps to limit the number of configurations that the chain can take. Further limitation of the number of pos-

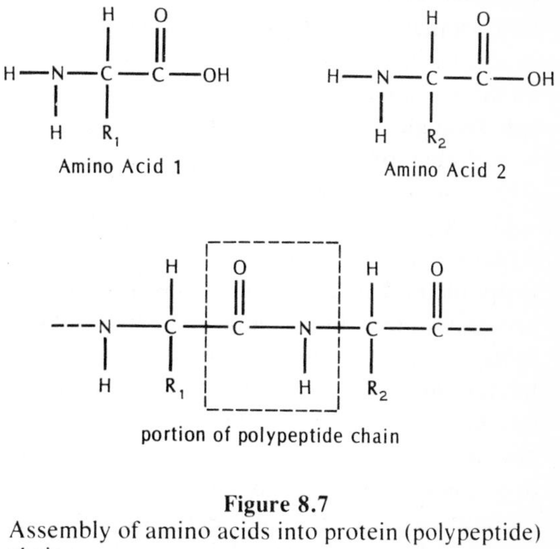

Amino Acid 1   Amino Acid 2

portion of polypeptide chain

Figure 8.7
Assembly of amino acids into protein (polypeptide) chains

Figure 8.6
Structure of an amino acid

The subunits or monomers of a protein are termed amino acids. An amino acid consists of a central carbon atom, called the alpha car-

sible configurations is provided by the considerable size of some of the R groups, since at some backbone chain positions the R groups would get in each other's way, i.e. they would exhibit steric hindrance.

The R groups include some as small as a single hydrogen atom (as in glycine) a long chain (such as $-CH_2-CH_2-S-CH_3$ in methionine) or a complex ring structure as in tryptophan. Some are of a hydrocarbon nature and consequently are hydrophobic (as in valine $-CH-(CH_3)_2$). Others with hydroxyl groups (as in serine $-CH_2-OH$) or amide groups (as in glutamine $-CH_2-CH_2CONH_2$) can participate in hydrogen bonds or dipolar interactions. Still others may be charged (as in glutamic acid $-CH_2-CH_2-CO_2^-$) and participate in electrostatic interactions.

For a particular protein, a particular sequence (called the primary structure) is specified by the gene responsible for the eventual assembly of that protein. Once formed, the polymer adopts secondary and tertiary structures which are dictated by the interactions between its R groups, and between R groups and solvent within the limits of the geometrical constraints already mentioned. One possibility is that the chain might first wind itself up into a helix (the secondary structure) and the helix would wind itself up into a more or less spherical globule (the tertiary structure). This, in a rather over-simplified manner, is essentially what happens with many proteins.

Of considerable importance to the final structure are the relative numbers of hydrophilic and hydrophobic R groups. If there are many more hydrophilic than hydrophobic groups, the molecule will have a rather extended shape—perhaps like a prolate or oblate ellipsoid—as the many hydrophilic groups are attracted to the solvent and the hydrophobic groups are repelled away from it and attempt to go to the centre of the molecule. As the relative number of hydrophilic groups

decreases the molecule will become spherical to reduce the surface area to volume ratio. In some cases, there are insufficient hydrophilic groups to cover even the surface of a sphere of the hydrophobic residues. In this case, clusters of molecules form. The surface area to volume ratio is then decreased relative to that of the individual molecules and the hydrophilic residues can then cover the surface of the cluster. The protein, insulin, clusters in groups of four molecules for this reason.

A number of secondary structures are known to exist. The first one discovered was the alpha helix in which the backbone of the polypeptide winds into a helix of diameter 0.36 nm and pitch 0.544 nm with 3.6 R groups per turn. The helix is stabilized by hydrogen bonds which form between the carboxyl oxygen of one amino acid and the hydrogen attached to the nitrogen atom of the fourth amino acid along the chain. The R groups protrude and, depending on their nature, prevail upon their particular segment of helix to be near the face or the interior of the molecule. Some R groups are particularly partial to forming α helices, others such as proline are helix breakers, and still others are rather impartial. Some proteins have considerable lengths of alpha helix but others have little or none. The structure of the alpha helix is shown in Figure 8-8. Alpha helices are particularly prevalent in the keratins which form the protective hair, wool, nails, feathers, etc. of vertebrates.

Other secondary protein structures include triple helices as in collagen, the major tensile component of connective tissue, the random or statistical coil as in portions of many enzymes, and the antiparallel pleated sheet of silk. These are shown in great detail in elegant drawings in the book by Dickerson and Geis. Many enzyme and carrier molecules will exhibit a mixture of two or three types of secondary structure.

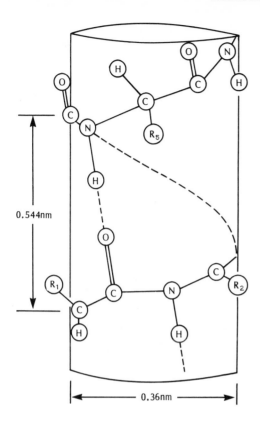

0.544nm

0.36nm

Figure 8.8
Diagrammatic representation of a portion of alpha helix

The secondary structure, once formed, will then be oriented in an apparently random but very specific configuration which must be exactly right if the protein is to function normally. Sometimes a genetic accident, called a mutation, can cause one R group to be substituted for another. This can, in some instances, cause just enough change in the overall structure of the protein to prevent it from carrying out its normal function with possibly fatal consequences.

The final configuration will be dictated by hydrophobic and hydrophilic interactions, as already discussed, plus a variety of electrosta-

tic, dipolar, and hydrogen bond interactions between different parts of the molecules. Thus, if a charged group is accidently replaced by a neutral group, the final configuration may be different.

Why is final configuration so important? Consider, for instance, those proteins that are enzymes. Each different enzyme has its own unique role—usually to "operate" on some substrate molecule to modify it or cleave it in some way. To accomplish this the substrate must first fit, by some combination of the interactions that have been discussed, to a particular location or active site on the enzyme after which the "operation" occurs. The fit is a very specific lock-and-key type coupling and any mutation that alters the effective shape of the active site will prevent the enzyme from functioning normally by destroying the fit between enzyme and substrate.

There is, of course, tremendous variability in protein structure which stems from the fact that if a typical protein of about 400 monomers is to be assembled from 20 different amino acids then there are $(20)^{400} = 10^{520}$ different possible sequences.

To this point, this section has probably left the student with the impression that all proteins are in a watery environment. The most thoroughly studied and best understood proteins are from a watery environment but there is an important, though less well understood group of proteins, which are partly or completely confined to membranes. Membranes have a hydrocarbon, and hence hydrophobic, interior (Figure 5-5). Consequently many membrane proteins have a structure that is "inside-out" with a hydrophobic exterior to interact with the membrane interior and a hydrophilic core. There is good evidence to suggest that for at least some membrane components the hydrophilic core may be arranged to provide an open pore from one side of the membrane to the other and thus facilitate the

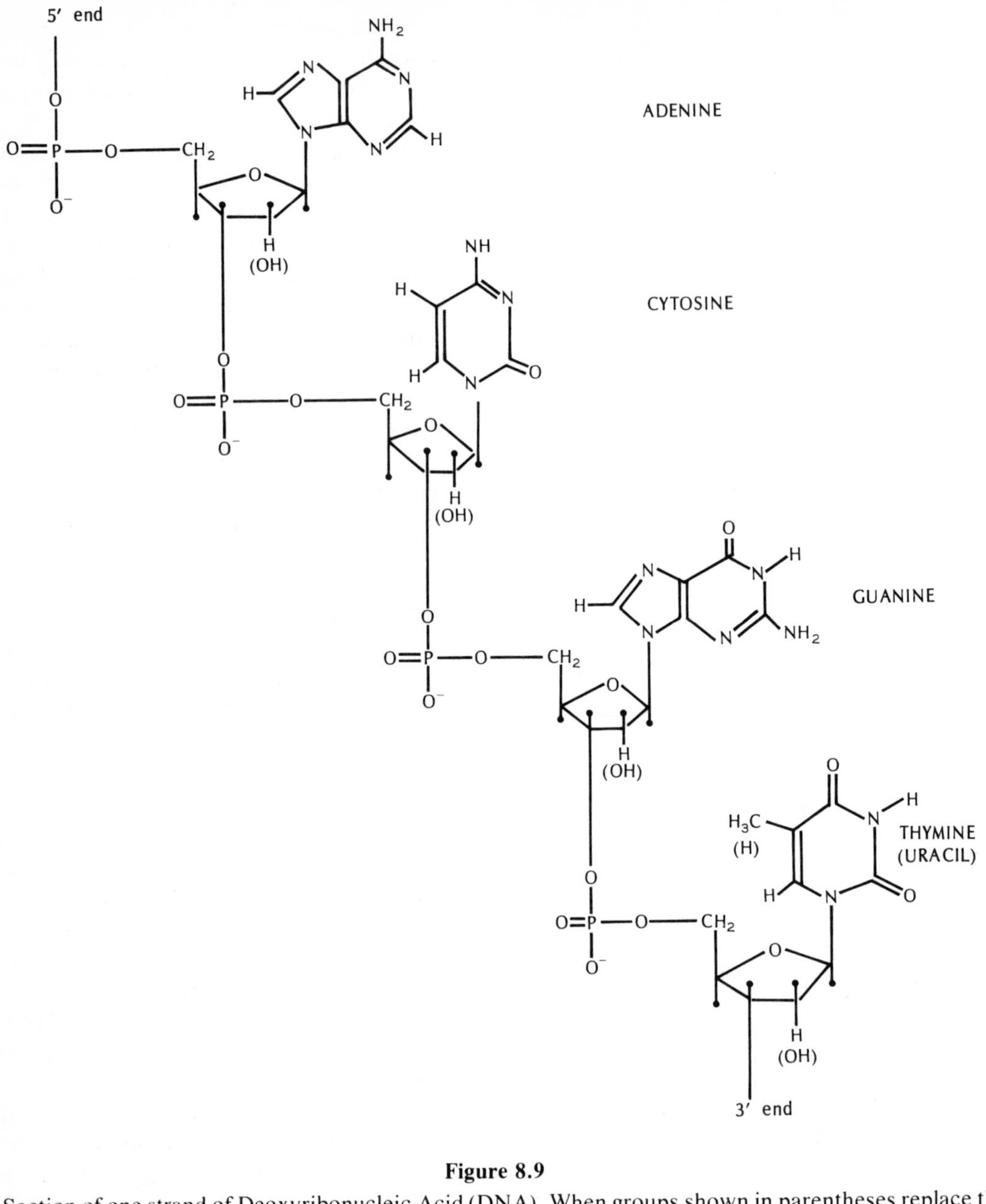

Figure 8.9

Section of one strand of Deoxyribonucleic Acid (DNA). When groups shown in parentheses replace the group immediately above them, RNA (Ribonucleic acid) results. In RNA the base uracil "replaces" the thymine of DNA. The 3' end connects to the number 3 carbon on the ribose sugar ring – as determined by a universal numbering convention

movement of water soluble entities across the membrane.

It has been stated that the primary sequence is controlled by a gene. The genes are located in the chromosomes where the information for amino acid sequences is stored in coded form usually in a length of DNA (deoxyribonucleic acid) molecule. One of the scientific triumphs of this century has been the breaking of this code and the understanding of the mechanisms by which DNA replication and protein formation is carried out and regulated. There is, unfortunately, no room in this volume for the details of this story and the student is referred to the excellent book "Molecular Biology of the Gene" by Watson.

The actual structure of DNA will however be considered as this structure was determined by the application of basic principles of physics, some of which will be discussed in the next section.

DNA is a two-stranded polymer made of sequences of four different monomers. Each monomer consists of a ribose sugar group, a phosphate group and one of four bases—adenine, cytosine, guanine, and thymine. The sugar-phosphate portions are linked together to form a backbone with the bases protruding from the chain. A section of one strand of DNA is shown in Figure 8-9 where the changes necessary to convert to RNA (ribonucleic acid) are also indicated. Two DNA chains normally are associated in a double helix in which cross-links between chains consist of hydrogen-bonded base pairs. The fitting is so precise that only two pairs can normally exist. These are guanine-cytosine and adenine-thymine. Thus if the base sequence on one chain is specified, the base-pairing mechanism automatically determines the sequence on the other chain. As indicated in Figure 8-10, the double helix has diameter of 2.0 nm and a length per turn of 3.4 nm with 10 base pairs per turn. The two backbones are assembled in opposite directions i.e. if the 3' end is at the bottom of one chain and the 5'

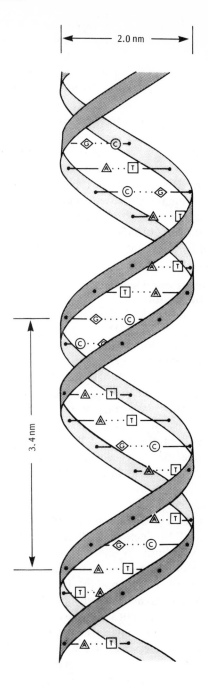

Figure 8.10
Structure of the DNA double helix (From: R. E. Dickson and I. Geis, The Structure and Action of Proteins, Harper and Row, 1969.)

end (see Figure 8-9 for explanation of 3′ and 5′) at the top, the reverse applies to the other chain.

RNA is formed, in most instances, against one of the DNA chains. Three types of RNA are formed but none has extensive regions of double helix. All in fact are single chains but in some cases short sections of double helix can form as a portion of a chain loops back and forms a double helix with itself. The different RNA molecules have the jobs of transporting the coded information to the protein assembly site, forming the site and bringing animo acids to the site. Again the reader is encouraged to consult Watson's book for full details.

8-4 Structure Determination and X-ray Diffraction

As stressed in the last section, the configuration of protein and DNA molecules is closely related to the function of the molecule. Thus it is important to learn about the structure of these molecules. A number of different methods are used to acquire such knowledge, but the most important has been the technique of x-ray diffraction. The perfection of this technique to the point where scientists can determine the exact position of each of the thousands of atoms in a protein molecule has been a major intellectual and technological achievement of man.

How does one go about looking at a protein molecule? In Section 3-10 it is shown that there is a limit to the resolution of detail which can be achieved by a microscope using visible light. This is not the fault of the microscope but is an inherent property of the wave nature of light. The best that any microscope can do is to form separate, distinct images of objects which are separated by at least one-half of the wavelength of the light used in the microscope. Objects closer than this are not resolved as separate points in the image.

In a molecule, the atoms are about 0.1 nm apart. Therefore if electromagnetic waves are used to examine a molecule with the objective of determining the position of the atoms, a wavelength of about 0.1 nm must be used. This is the x-ray region of the electromagnetic spectrum. In short, an x-ray microscope is required.

Unfortunately it has been impossible to construct an x-ray microscope because there is no simple way to focus x-rays. Thus one cannot directly use x-rays to produce magnified images of molecules. Instead the more indirect technique of x-ray diffraction is used.

The term diffraction refers to a phenomenon common to all types of waves in which a travelling wave is scattered by objects with which it interacts and the scattered waves subsequently overlap or are superimposed in the surrounding space to produce variations in wave intensity. The process is described in detail for x-rays in the following paragraphs.

An x-ray beam may, for some purposes, be considered as a beam of high-energy photons and for other purposes as a beam of electromagnetic waves. The latter model is the most useful for describing diffraction phenomena.

Figure 8-11 illustrates a typical x-ray diffraction experiment. A simple molecule consisting of only two atoms (points labelled 1 and 2) is shown irradiated by an x-ray beam incident from the left. The actual x-ray source is not shown in the diagram and the details of how x-rays are produced are not important here. The apparatus is usually designed to produce an x-ray beam which is (i) monochromatic, i.e. contains essentially only one wavelength and (ii) highly collimated, i.e. the x-ray wavefronts are planar or equivalent, the rays are straight, parallel lines. These points are illustrated in Figure 8-11 where ab represents a plane x-ray wavefront arriving at atom 1. The wavefront represents a "crest" or re-

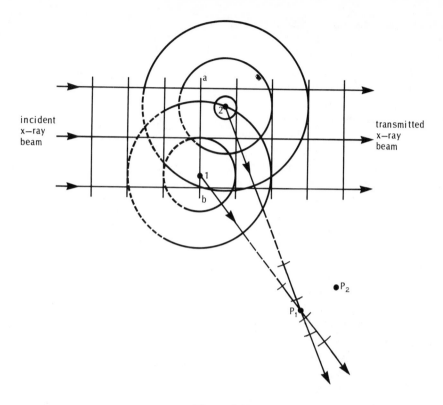

Figure 8.11

Spherical waves scattered from atoms 1 and 2 irradiated by a beam of x-rays

gion of maximum electric field in the wave.

The incident electromagnetic wave sets up an oscillating electric field at each atom. This causes the electrons in the atom to oscillate with the same frequency as the x-ray wave. The energy required for the oscillation comes from the incident x-ray beam. The oscillating electrons in turn act as sources of electromagnetic waves and reradiate their acquired energy in all directions. Thus each atom acts as a source of reradiated spherical waves of the same frequency and wavelength as the original incident x-rays. The whole process is called "scattering." The x-ray scattering ability of an atom depends on the number of electrons in the atom. The amplitude of the

scattered spherical waves is not equal in all directions but is largest in the "forward" direction i.e. the direction of the transmitted beam, decreasing to almost zero in the opposite direction. Figure 8-11 shows the scattered spherical waves radiating out from atoms 1 and 2. The two scattered waves are superimposed at all points in the space surrounding the molecule.

It is evident that, in general, the scattered waves from the different atoms in a molecule will not start out "in phase" with each other because of the different distances travelled by the incident wavefronts to reach each atom. Thus in Figure 8-11, atom 2 is further from the x-ray source than is atom 1 so each incident wavefront (such as ab) reaches 2 slightly later than it reaches atom 1. Hence the scattered wavefronts from atom 2 will lag in time slightly behind those from atom 1.

Now consider the two scattered x-ray wave trains as they simultaneously pass through a typical point P_1 some distance from atoms 1 and 2. In general the two waves will not have the same amplitude since this depends on the electron content of the two source atoms (which usually will be different) and also on the different distances from atoms 1 and 2 to P_1. (The amplitude of a spherical wave decreases with distance from its source.) In addition the two waves reaching P_1 will in general have different phase angles, that is (normally) the two waves will not produce oscillations at P_1 which are "in step" with one another (see Appendix 3). This is because (i) the waves did not start out in step at the source atoms as explained in the preceding paragraph and (ii) there is a difference in distance from atoms 1 and 2 to P_1. Even if the waves started out at atoms 1 and 2 in step with one another, they would not reach P_1 in step unless the distances from 1 and 2 to P_1 happened to be equal or happened to differ by an integral number of wavelengths of the scattered waves. It is evident that the amplitude and phase of the two scattered waves at P_1 depends on the nature and position of the two atoms in the scattering molecule as well as the distance of P_1 from the centre of the scattering molecule.

Each scattered wave, if it passed through P_1 by itself, would produce an oscillating electric field. When the two scattered waves simultaneously pass through P_1, the resultant electric field at each instant is simply the vector sum of the fields each would produce alone. The resultant is itself an oscillating electric field of the same frequency as the two scattered waves, as shown in Figure 8-12(a) and (b).

The resultant oscillation at P_1 will have two

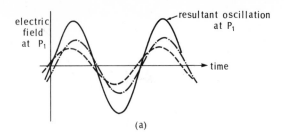

(a)

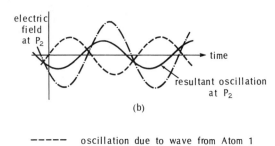

(b)

- - - - -   oscillation due to wave from Atom 1

— · — · —   oscillation due to wave from Atom 2

Figure 8.12
a) The electric field oscillations at point P_1 where the waves from atoms 1 and 2 are almost in phase.
b) The electric field at a second point P_2 where the waves from atoms 1 and 2 are nearly 180° out of phase

important properties: (i) amplitude and (ii) phase, both of which depend on the amplitudes and phases of the scattered waves reaching P_1. Figure 8-12(a) illustrates the case where the distances from atoms 1 and 2 to P_1 have values which cause the two scattered waves at P_1 to be almost "in phase" with each other. In this case the resultant oscillation has a large amplitude. Figure 8-12(b) illustrates the case for a different point (P_2) where the distances from atoms 1 and 2 to P_2 are such that the two scattered waves are almost 180° out of phase at P_2. The resultant oscillation at P_2 has a very small amplitude. In addition to differing in amplitude, the resultant oscillations at P_1 and P_2 do not reach their maxima together i.e. they differ in phase.

In general the resultant oscillations at P_1 and P_2 differ in both amplitude and phase depending on the amplitudes and phases of the scattered waves reaching these points. The latter depends on the nature and position of the atoms in the scattering molecule. Thus ultimately the amplitude and phase of the resultant oscillation at points such as P_1 and P_2 depend on the structure (kinds and arrangement of atoms) of the scattering molecule. Although this example has considered only two atoms in the molecule, the conclusions are valid for any number of atoms. The resultant amplitude and phase would vary from point to point in a manner dependent on the molecular structure.

A complete analysis of the scattering shows that if one could measure the amplitude and phase of the resultant scattered electromagnetic wave in each direction from the scattering molecule and for various incident directions and wavelengths, one could calculate the structure of the scattering molecule. This is the basic principle of the x-ray diffraction technique.

There is one fundamental problem in x-ray diffraction. As indicated above, two quantities are required (i) the amplitude and (ii) the phase of the resultant wave. The amplitude can be measured relatively easily by measuring the intensity of the scattered x-rays in the various directions using standard detectors such as x-ray sensitive film or ionization counters. The amplitude is proportional to the square root of the measured intensity. Unfortunately there is no direct way to measure the phase of the resultant wave. (This is called the "phase problem" in the diffraction technique.) A great deal of ingenuity has been used to devise ways to determine the phase information by indirect means.

Of course, a single molecule cannot be used as a scattering sample. Many molecules must be used to obtain measurable scattering and to be able to physically handle the sample. Further, the molecules must all be oriented as far as possible in an identical manner. The ideal sample for molecule structure determination is therefore a single crystal of the material. During the past two decades, a great deal of expertise has been developed in growing crystals of large protein molecules and aligning semicrystalline fibres of long molecules such as DNA. The procedures are difficult but, fortunately, small crystals (about 1 mm diameter or less) will suffice for x-ray diffraction studies.

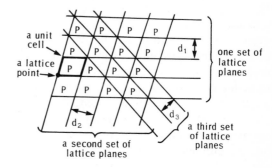

Figure 8.13
A two-dimensional crystalline array showing the repeating unit P and the imaginary lattice points and planes

When molecules are arranged in a crystalline solid they are arranged in a regularly repeating three-dimensional array. The repeating unit might be a single molecule or possibly a group of molecules. This is illustrated for two dimensions in Figure 8-13 where the letter P represents the repeating unit, possibly a protein molecule. It is useful to associate such a structure with a regular, repeating, array of imaginary points called a lattice, each repeating unit being associated with a lattice point. The lattice points mark the corners of a unit cell each one of which contains one of the repeating units. It is also possible to imagine many sets of parallel planes (lattice planes) passing through the lattice points as illustrated in Figure 8-13 which shows only three of the many possible sets. Each set of planes is characterized by the perpendicular distance (d_1, d_2, d_3 etc. in Figure 8-13) between adjacent planes of a set. These distances vary from one set of planes to another within a given lattice.

When a small single crystal is used in a diffraction experiment, the x-rays are scattered from all the atoms in all the molecules in the sample. Because the molecules are in a regular array, there is a very interesting result. If the crystal is placed in the x-ray beam with a random orientation, very little x-ray scattering will be observed in any direction. This is because the resultant waves from the molecules are out of phase with one another to such an extent that the vector sum of all the oscillating electric fields is essentially zero at all points. However if the crystal is rotated, certain orientations can be found at which an intense diffracted beam is produced in some specific direction. For these special orientations the scattered waves from the many molecules are "in phase" in the specific direction, giving a very intense diffracted beam in that direction.

In 1912, W. L. Bragg showed that the occurrence and direction of the intense diffracted beams from a crystal could be interpreted in a simple way. His reasoning and conclusions are outlined below.

Bragg first considered scattering from a single plane of regularly spaced atoms as in Figure 8-14. The incident x-ray beam, represented by parallel rays 1 and 2, and associated plane wavefronts make an angle θ between the rays and the lattice plane. The spherical scattered waves from two atoms are shown. Most of the energy continues in the transmitted beam. When the atoms are regularly spaced, the scattered waves are in phase with each other in only two directions. One of

these is the direction of the transmitted beam and this diffracted beam is not detectable. The other direction, producing a detectable beam, occurs in the direction with the following properties (illustrated by rays 1_s and 2_s in Figure 8-14): (i) the incident beam, the normal to the lattice plane, and the diffracted beam lie in a single plane and (ii) the angle r between the diffracted rays and the lattice plane is equal to the angle θ between the incident rays and the lattice plane. In all other directions the scattered waves are out of phase and yield zero resultant. The rules for the direction of the detectable diffracted beam are the same as the rules for the reflection of light from a mirror. Hence, Bragg's first conclusion was that x-rays are diffracted from a lattice plane as if they are simply reflected from the plane.

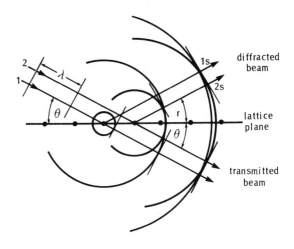

Figure 8.14
The scattering of x-rays from a single plane of regularly spaced atoms

However there is an important restriction on the angles θ at which x-rays can be "reflected" from a *set* of parallel lattice planes. This is due to the superposition of the diffracted beams from all the planes in the set. These beams are all parallel and hence are superimposed. For most angles of incidence θ, the diffracted beams are out of phase with each other to the point where the resultant amplitude of the overall "reflected" beam is zero and all of the scattered energy appears in the transmitted beam. However there are some angles for which the diffracted beams from the planes in a set will be "in phase" and hence produce a resultant reflected beam of large amplitude and intensity. These directions can be determined from Figure 8-15.

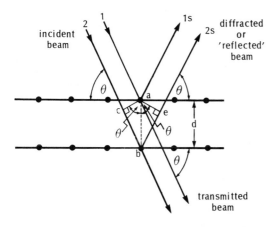

Figure 8.15
The scattering of x-rays from two adjacent planes from a set of parallel lattice planes, illustrating the geometry associated with the Bragg equation

Incident ray 1 scatters from atom a and ray 2 scatters from atom b directly below in the adjacent plane. The planes are separated by a perpendicular distance d. The distance $cb + be$ is the extra distance travelled by ray 2 in reflecting from the lower plane, over that travelled by ray 1. The reflected or scattered waves, represented by rays 1_s and 2_s will be "in phase" if:

$$cb + be = \lambda, \ 2\lambda, \ 3\lambda \ \text{etc.}$$

or

$$cb + be = n\lambda \quad \text{where } n = 1, 2, 3 \text{ etc.}$$

Now

$$cb = be = d \sin \theta$$

Therefore the above condition becomes:

$$2d \sin \theta = n\lambda \qquad n = 1, 2, 3, \ \ldots \quad (8\text{-}3)$$

Equation (8-3) is known as the Bragg equation and enables one to determine the angles of incidence θ, at which reflections will occur from a given set of lattice planes.

EXAMPLE

A set of lattice planes has a d-spacing of 0.40 nm. At what angles of incidence will Bragg reflections occur for x-rays of wavelength $\lambda = 0.16$ nm?
From the Bragg equation:

$$\sin \theta = \frac{n\lambda}{2d}$$

$$= \frac{n \times 0.16 \ \text{nm}}{2 \times 0.40 \ \text{nm}}$$

$$= n \times 0.20$$

Thus for: $n = 1$ (first order reflection)

$$\sin \theta = 1 \times 0.20 \text{ and } \theta = 11.5°$$

$n = 2$ (second order reflection)

$$\sin \theta = 2 \times 0.20$$

$$= 0.40 \text{ and } \theta = 23.6°$$

Similarly for $n = 3$, 4 and 5. For $n = 5$, $\sin \theta = 5 \times 0.20 = 1.0$ and $\theta = 90°$ the largest possible angle of incidence. Thus for this particular set of planes and for this particular wavelength there are five different angles of incidence or "Bragg angles" at which a strong diffracted or "reflected" beam will occur. At all other angles, no diffracted beam would be detected.

It should be noted, as shown in Figure 8-15 that when a Bragg reflection does occur, the angle between the "reflected" beam and the transmitted beam is 2θ, twice the angle of incidence.

Bragg's overall conclusion was that the diffracted beams from a crystal occur as if each were reflected from a set of parallel lattice planes subject to the restriction that a reflected beam will occur only for those angles of incidence given by the Bragg equation.

The Bragg conditions determine the directions in which diffracted beams may occur from a crystal and these are determined entirely by the wavelength used and the lattice spacings (d). The latter are in turn determined by the spacing of the repeat unit P in the crystal (and not by the arrangement of atoms within P). The Bragg conditions say nothing about the intensity of the possible reflections. The intensity of the reflections is determined by the structure of P, i.e. the kinds and positions of the atoms making up P. It is this information which is the objective of the whole measurement. A complete analysis shows that if there are N repeat units in the crystal, the amplitude of the wave in a given Bragg reflection is just N times the amplitude one would obtain in that same direction from one unit (and the intensity is N^2 times as great). The phase of the reflected wave is the same as would be obtained from a single repeat unit P. Thus the effect of the lattice or regular array of units is to "sample" the diffraction pattern of one unit in each of the directions at which a Bragg reflection occurs. By measuring the intensity of the various Bragg reflections, the scientist obtains the amplitude of the diffracted beam in that direction and by various indirect techniques obtains the phase of the wave. By determining these two parameters for a large number of reflections, the structure of the repeat unit may be determined. If the repeat unit is a complicated structure such as a protein molecule, several thousand reflections must be measured and millions of calculations performed on high-speed computers. As stated earlier, the analysis of the structure of macromolecules has been a major intellectual and technological achievement. A great deal of effort has been required but this effort has paid great dividends in our understanding of the fundamental processes in molecular biology.

Problems

1. (a) What is the electrical potential energy of a system consisting of one positively and one negatively charged molecule separated by 5 nm of water? Assume the charges are equal in magnitude to the charge on the electron.
   (b) What is the electrical potential energy if the water is removed?
2. The separation between molecules which are interacting via London dispersive forces is increased by 10 percent. Does the electrical potential energy increase or decrease? By what percentage?
3. A hydrophobic interaction refers to:
   (a) interactions between transient charges on molecules
   (b) the tendency of some molecules to become highly solvated in water
   (c) dipole–dipole, dipole-induced dipole forces
   (d) the tendency of some molecules to escape from a watery environment
   (e) none of the above
4. The amino acid alanine has a CH_3 group as its side chain. It is therefore chemically inert and its interactions are due to dispersive hydrophobic forces. The amino acid aspartate has a charged side chain. If the structure of any enzyme in an aqueous environment is carefully examined, which

of the following is correct?

(a) The alanine and asparate molecules would be generally buried deep within the molecule

(b) The alanine would be mostly at the surface and the aspartate mostly deep within the molecule

(c) Both molecular types would congregate at the surface

(d) The alanine would be predominately deep within the molecule while the aspartate would be generally at the surface

(e) Both molecules would be randomly located and no generalizations could be made

5. (a) A beam of x-rays of wavelength 0.070 nm is used to determine the structure of a protein. One of the sets of parallel lattice planes in the protein crystal has a separation (d) of 5.0 nm. What are the angles of incidence (between the x-ray beam and the planes) for the first three orders of Bragg reflections from this set of planes?

(b) If the protein crystal in (a) is oriented so as to produce the first order reflection from the above planes, what will be the angle between the beam transmitted through the crystal and the diffracted or "reflected" beam?

6. (a) A polychromatic beam of x-rays (one containing a distribution of wavelengths) contains all wavelengths from 0.080 nm to 0.30 nm. The beam strikes a crystal at an angle of incidence of 30° to a set of crystal planes with a spacing (d) of 0.40 nm. What wavelengths will be "reflected" from the beam by this set of planes?

(b) At what angle, relative to the transmitted x-ray beam, will the above reflected x-rays emerge from the crystal?

Chapter 9
Properties of Fluids

9-1 Introduction

Fluids (liquids and gases) are important in determining the internal and external environment of living systems. A liquid provides the medium in which the building blocks of life, the amino acids, are brought together for synthesis into proteins. Liquids carry nutrients to the various cells to fuel the metabolic processes. Air, a gas, provides life-sustaining oxygen for metabolic processes and pressure to keep organisms in their proper shape. A knowledge of the basic properties of fluids such as pressure, surface tension, and viscosity is necessary to an understanding of how living systems function.

9-2 Pressure

Within any fluid, whether it be liquid or gas, there will be pressure that increases with increasing depth. This pressure is defined as the magnitude of the normal force per unit surface area. The pressure will depend not only on the depth of fluid or fluids involved but also on the actual weight per unit volume of the fluid which is determined by the density of the fluid and the acceleration due to gravity. Density (usually represented by the Greek letter rho, ρ) is measured as mass per unit volume. For any given fluid there will be some variation of density with changes in either pressure or temperature. For liquids, these changes are usually so slight that, within the limits required for life processes, they can be ignored unless extreme precision is required. For example, if one m^3 of water at 4° C were subjected to twice the normal atmospheric pressure its volume would decrease from $1.00 \, m^3$ to $0.99995 \, m^3$ and its density would increase to $1000.05 \, kg \, m^{-3}$. A gas, on the other hand, subjected to the same pressure change, would have its volume halved and its density doubled. The density of the gas would also change markedly with temperature.

To evaluate the variation of pressure with depth in a fluid at rest, consider Figure 9-1.

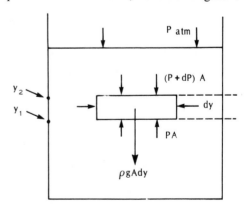

Figure 9.1

Forces acting on an element of fluid at rest

There is indicated an element of fluid of surface area A and thickness dy. Pressure has already been defined as the normal force per unit surface area. Therefore there is a force of magnitude PA exerted on the lower surface of the element of fluid and a force $(P+dP)A$ on the upper surface. The directions of these forces are as indicated in the diagram. There is also a force $\rho A\, dy$, the weight of the element, acting down. If the fluid element is at rest, then the resultant vertical force must be zero (by Newton's second law). That is

$$PA - (P+dP)A - \rho gA\, dy = 0$$

or

$$dP = -\rho g\, dy \qquad (9\text{-}1)$$

Let us consider in some detail the pressure variations with depth in (i) liquids (ii) gases.

(i) Liquids

We have seen that density variation with pressure in a liquid is negligible. Therefore ρ is a constant and the expression Equation (9-1) can be integrated to yield

$$\int_{P_1}^{P_2} dp = -\rho g \int_{y_1}^{y_2} dy$$

$$P_2 - P_1 = -\rho g (y_2 - y_1)$$

or

$$P_1 = P_2 + \rho g (y_2 - y_1) \qquad (9\text{-}2)$$

Note that the shape of the container does not affect the pressure and that the pressure is the same at all points at the same depth.

If y_2 is taken as the surface of the liquid then $P_2 = P_{\text{atm}}$ and the absolute pressure at any depth is

$$P = P_{\text{atm}} + \rho g h \qquad (9\text{-}3)$$

where h is the depth below the surface at which the pressure is measured. Note that the pressure due to the atmosphere is transmitted through the liquid, that is, pressures are additive. The pressure difference $(P - P_{\text{atm}})$ is

termed the gauge pressure at depth h and arises from the weight of the liquid above the point.

Dimensional analysis shows that the dimensions of pressure are

$$[P] = \frac{\text{force}}{\text{area}} = \frac{ML}{T^2 L^2} = ML^{-1}T^{-2}$$

The units of pressure are $N\,m^{-2}$ or pascals, abbreviated Pa. Pressure must always be expressed in terms of force per unit area when calculations are made involving physical formulae. However, it is frequently expressed in other, less ideal units which usually have to be converted before calculations can be made. For example, the pressure may be quoted in mm of mercury. This value is converted to a physical pressure unit by multiplying it by the density of mercury and the acceleration due to gravity i.e. $P = \rho g h$. Other commonly encountered pressure units are an atmosphere (10^5 Pa), a bar (10^5 Pa), a torr (1.3×10^2 Pa) etc.

Most pressure gauges register the difference in pressure between two points. This difference (which can be positive or negative) is a gauge pressure and will not be an absolute pressure unless (as rarely happens) one of the two points is at zero pressure. Consider, for example, an automobile tire which is frequently filled to a pressure of approximately 2 atmospheres. This means that the pressure inside is 2 atmospheres more than outside and thus is a gauge pressure. Since the pressure outside is one atmosphere then the total or absolute pressure inside will be 3 atmospheres. One commonly used pressure-measuring device is the manometer. Suppose that on a day when the atmospheric pressure is 760 mm Hg, a mercury U-tube manometer (Figure 9-2) is connected to a chamber of some gas. The manometer indicates a difference of 190 mm Hg (gauge pressure). Note that in this case the level of the liquid is higher on the chamber side of the manometer than on the side open to the atmosphere. This indicates that the pressure P inside the chamber is less than atmospheric pressure P_{atm}. Since 190 mm Hg $= \frac{1}{4}$(760 mm Hg), P is less than P_{atm} by $\frac{1}{4}P_{atm}$. The absolute pressure inside the chamber then is 3/4 of an atmosphere or 0.75×10^5 Pa. In addition to mercury, oil and water are frequently used as manometer liquids. Remember that, in all cases, if the gauge pressure is quoted in terms of a height of some liquid then before using it in any calculations it must be converted to a physical pressure unit using the expression $P = \rho g h$, where ρ is the density of the liquid.

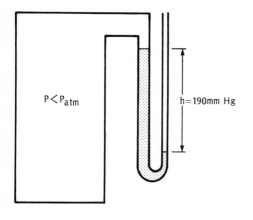

Figure 9.2
Manometer

EXAMPLE: PRESSURE IN WATER

The gauge pressure at a depth of 1 metre would be

$$\rho g h = 10^3 \text{ kg m}^{-3} \times 9.8 \text{ m s}^{-2} \times 1 \text{ m} = 10^4 \text{ Pa}$$

or $\frac{1}{10}$ of normal atmospheric pressure (10^5 Pa). The total pressure at this depth would, of course, be 1.1×10^5 Pa or 1.1 atmospheres.

A swimmer becomes very much aware of this kind of increase in pressure when sub-

merging as little as 2 metres since the pressure difference across the eardrum increases from zero at the surface to a significant amount rather quickly. For example, if the pressure on the inside of the eardrum were 10^5 Pa and on the outside 1.2×10^5 Pa then, since the area is approximately 0.66×10^{-4} m^2, there would be a net force on eardrum of

$$F = PA$$
$$= 0.2 \times 10^5 \times 0.66 \times 10^{-4}$$
$$= 1.3 \text{ N}$$

This is a lot of force, in fact, ten times that required to rupture the eardrum. The eardrum does not rupture since the pressure inside the eardrum is also increased due to the water pressure acting over the body and to involuntary contraction of muscles in the respiratory system.

EXAMPLE: THE HYDROSTATIC FACTOR IN THE HUMAN CIRCULATORY SYSTEM

The change of pressure with depth of a liquid is called the hydrostatic factor and is important in the vascular system of the human body. At the heart level, the arterial pressure is 100 mm Hg. When the body is horizontal the average arterial pressures in the brain and feet are approximately the same as at the heart namely 100 mm Hg because they are all at the same level. If the body is erect, the hydrostatic factor reduces the arterial pressure in the brain and increases that in the feet. Suppose the head is 0.50 m above the heart. Then the arterial pressure in the head is

$$P_{head} = 100 - \frac{\rho(\text{blood}) h}{\rho(\text{mercury})}$$
$$= 100 - \frac{(1050)}{13,600}(500)$$
$$= 60 \text{ mm Hg}$$

If the feet are 1.30 m below the heart the arterial pressure in the feet is

$$P_{\text{feet}} = 100 + \frac{\rho(\text{blood})\,h}{\rho(\text{mercury})}$$

$$= 100 + \frac{1050}{13{,}600}(1300)$$

$$= 200 \text{ mm Hg}$$

Thus the cause of swollen feet and the reason for lowering the head as a cure for fainting should be evident.

(ii) Gases

Let us consider the atmosphere as our example. Analysis of the atmosphere on a dry basis reveals it to consist of 78 percent N_2, 20 percent O_2, 0.9 percent argon, 0.03 percent CO_2 and small amounts of xenon and krypton.

The atmospheric envelope being compressible and influenced by the gravitational field of the earth is relatively dense near the surface and the density decreases rapidly in a non-linear fashion with increasing altitude. Normal sea level pressure is taken to be 0.76 m of mercury but fluctuations of ±0.04 m are not uncommon. For the first few thousand feet above sea level, the pressure will drop about 0.025 m Hg per 300 m of height. The atmosphere is usually divided into the troposphere (0–12,000 m), the stratosphere (12,000–26,000 m) and still higher regions. At 100 km, pressures are as low as in a good vacuum tube.

Almost all living organisms on earth above sea level exist at pressures between 0.35 and 0.76 m of mercury. At any altitude where the pressure is less than 0.52 m of mercury (≈ 3000 m above sea level) exertion is likely to result in distress. Breathing pure oxygen will help but not indefinitely. At 14,000 m (0.13 m of mercury) even with 100 percent oxygen, there will be barely enough oxygen crossing the alveolar walls in the lungs to maintain consciousness and beyond this elevation pressurized systems are required. At 21,000 m, the pressure is the same as that of the water vapour in the body. Any fluid will

boil when its vapour pressure is the same as the prevailing atmospheric pressure so that actual boiling of body fluids would occur at this elevation in an unpressurized system and death would occur very rapidly.

At and near normal temperature and pressure most gases and gas mixtures obey the perfect gas law reasonably well. This law can be derived from first principles via the kinetic theory of gases or by combining Boyle's and Charles' laws and is usually stated

$$PV = nRT \qquad (9\text{-}4)$$

where P is the absolute pressure (Pa), V the volume (m^3), n the number of moles of gas, T the temperature (kelvin $= 273 + °C$), R is the universal gas constant $= 8.3147 \text{ J K}^{-1} \text{ mol}^{-1}$. Since 1 mole of a gas contains Avogadro's number ($N_A = 6.02 \times 10^{23}$) molecules of gas then the universal gas constant per molecule would be:

$$\frac{R}{N_A} = 1.38 \times 10^{-23} \text{ J K}^{-1} \text{ (molecule)}^{-1}$$

$$= k \text{ (Boltzmann's constant)}$$

The density of a gas will be $\rho = nMV^{-1}$ where M is the molecular weight of the gas. From the perfect gas law then

$$\rho = \frac{nM}{V} = \frac{PM}{RT} = \frac{PM}{kN_A T} = \frac{Pm}{kT} \qquad (9\text{-}5)$$

where the mass per molecule $m = M(N_A)^{-1}$.

The variation of pressure with distance is given by Equation (9-1) as

$$dP = -\rho g\, dy$$

Employing Equation (9-5),

$$\frac{dP}{P} = \frac{-mg}{kT}\, dy \qquad (9\text{-}6)$$

This equation can be integrated from sea level, h_1, (where the pressure is P_o) to any elevation h_2 (where the pressure is P). To simplify the calculation we will assume that g and T are constant as we go up. The temperature is reasonably constant within the pressure range of physiological interest and the decrease of g is quite small over the associated change of altitude. The integration then becomes

$$\int_{P_o}^{P} \frac{dP}{P} = \frac{-mg}{kT} \int_{h_1}^{h_2} dy$$

and the solution is:

$$ln \frac{P}{P_o} = - \frac{mg(h_2 - h_1)}{kT}$$

or

$$P = P_o \exp\left[-mg(h_2 - h_1)/kt\right] \quad (9\text{-}7)$$

Equation (9-7) is known as the Barometric Formula and indicates an exponential decrease in pressure with elevation. The term $mg(h_2 - h_1)$ in the exponent is the gravitational potential energy of an air molecule. The product kT (sometimes called the thermal energy) is also an energy term so that the exponent, as it must be, is a dimensionless quantity.

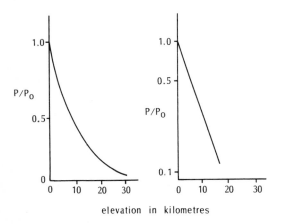

Figure 9.3
Variation of pressure with height in the atmosphere (linear and semilogarithmic plots)

A graph of P/P_o vs $h_2 - h_1$ is shown on linear and semilogarithmic scales in Figure 9-3.

9-3   Surface Tension

Small drops of water on wax paper almost form themselves into spheres and would do so if it were not for the force of gravity. If you have carefully watched films of the astronauts in space where the influence of gravity was negligible you no doubt saw perfect spheres of water floating around the inside of the capsule. The sphere is the shape with least surface area for a given volume. Thus the formation of spherical drops seems to indicate that the surface of the liquid has a higher energy than the bulk and is trying to achieve a state of minimum energy by reducing its surface. This excess surface energy is called the surface tension, γ. Surface tension effects are very important in a number of biological phenomena including the rise of water in plant stems, the expansion of the lung alveoli in breathing, and the walking on water of many water bugs.

Since water surfaces tend to behave almost like a skin, some authors have drawn an analogy between surface tension and a rubber sheet. This is, unfortunately, a very bad analogy since when the rubber sheet is stretched no new surface is created, the old is merely expanded, and each successive increment of surface area requires a greater force than the one before it. When a liquid surface is extended, it is done by creating new surface from molecules drawn from the bulk of the liquid. Thus, each successive increment in surface area requires the same force as the one before and the process can continue until there are no more interior molecules with which to create new surface. The interior molecules are attracted equally in all directions by their neighbours and thus feel no net force. Surface molecules are attracted by molecules beside them and by molecules within the bulk liquid but not by molecules from outside. Hence, work has to be done to pull them from the bulk, where they experience no force, to the surface, where they experience an attractive force back into the bulk.

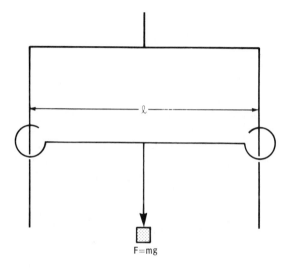

Figure 9.4
Apparatus for the definition of surface tension

We have defined surface tension as an excess surface energy. An equivalent but perhaps more useful definition is in terms of the forces present; that is, associated with every surface there is a surface tension which is simply the ratio of the surface force to the length along which the force acts (the force being perpendicular to the length). Consider the following. If a wire frame with a moveable cross-bar is dipped in a soap solution a film will be formed which can support a small mass m (for a particular position of the cross-bar), Figure 9-4. Within limits the cross-bar can then be moved up or down and will remain at rest when released indicating that there is present an upward force which is independent of the area of the film. This force must be equal to

the downward force $F = mg$ (for convenience we assume the cross-bar to be of negligible mass). The soap film has two surfaces (front and back with water in between). The upward force acts along the full length of the bar. This force is then $2\gamma\ell$. Since the forces balance (evidenced by the fact that the bar remains at rest) then

$$2\gamma\ell = F$$

or (9-8)

$$\gamma = F/2\ell$$

Surface tension γ has dimensions of $MLT^{-2}L^{-1}$ which is a force per unit length or equivalently an energy per unit area and units $N\,m^{-1}$ or $J\,m^{-2}$.

A wide variety of values of surface tension exist in nature. Table 9-1 gives a few values at 20°C.

Table 9-1
Surface Tension for Various Liquids

Liquid	Surface Tension
Water	$72.8 \times 10^{-3}\,N\,m^{-1}$
Blood plasma	$50 \times 10^{-3}\,N\,m^{-1}$
Lung surfactant	$1 \times 10^{-3}\,N\,m^{-1}$
Benzene	$28.9 \times 10^{-3}\,N\,m^{-1}$
Mercury	$464 \times 10^{-3}\,N\,m^{-1}$

Not surprisingly, surface tension is quite temperature-dependent; for water this effect is indicated in Table 9-2.

Table 9-2
Surface Tension of Water at Various Temperatures

Temperatures	Surface Tension
0°C	$75.6 \times 10^{-3}\,N\,m^{-1}$
20°C	$72.8 \times 10^{-3}\,N\,m^{-1}$
60°C	$66.2 \times 10^{-3}\,N\,m^{-1}$
100°C	$58.9 \times 10^{-3}\,N\,m^{-1}$

These values are obtained only with very pure water in very clean apparatus. Very small amounts of material in solution can cause considerable change. The addition of 3 molecules of acetic acid, for example, to 1000 molecules of water will decrease the surface tension by approximately 4 percent. Some materials decrease the surface tension (e.g. ammonium hydroxide) while others (e.g. potassium hydroxide) increase it.

Interesting results of biological significance occur when we examine the effects of surface tension on curved surfaces. Consider a soap bubble. We know we have to blow to create one so the pressure must be greater inside than out. The bubble, however, can achieve an equilibrium size when the surface tension is sufficient to counteract the force exerted by the greater interior pressure.

Figure 9-5 shows one-half of a soap bubble. Half bubbles can be created by blowing bubbles via a wire frame (shades of your childhood). P is the internal absolute pressure. P_{atm} is the atmospheric pressure. The excess internal pressure, $P - P_{atm}$, tries to separate the two halves; that is

$$P - P_{atm} = \frac{F}{A} \qquad (9\text{-}9)$$

where A is the projected area πr^2 and F is the force trying to move the half shown to the right. From Equation (9-9) $F = (P - P_{atm})\pi r^2$. To counteract this force, the surface tension exerts an attractive force to the left all around the "cut" edge as it is attracted to the other half. This attractive force is $F = \gamma \ell$ where ℓ is the length along which the force acts (the circumference $2\pi r$) which must be doubled since the bubble has two surfaces (inner and outer). Hence $F = 4\pi\gamma r$. The two forces are equal if the bubble is in equilibrium and then

$$(P - P_{atm})\pi r^2 = 4\pi\gamma r$$
$$(P - P_{atm}) = \frac{4\gamma}{r} \qquad (9\text{-}10)$$

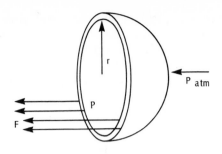

Figure 9.5
Half soap bubble

This indicates the surprising property of (soap) bubbles that the smaller the radius the bigger the internal pressure. If one blows two bubbles and connects them (not an easy task since the first one usually breaks about the time you have the second one) the smaller bubble will "deflate" and increase the size of the larger one as indicated in Figure 9-6.

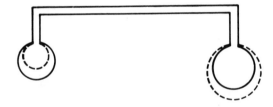

Figure 9.6
Effect of joining two soap bubbles of unequal radius. The initial size is indicated by the solid line; the final size by the broken line

If a bubble of air floating upwards in a beaker of water were subjected to the same type of analysis as the soap bubble, a similar but not identical equation would be obtained because this bubble has only one surface.

$$\Delta P = \frac{2\gamma}{r} \qquad (9\text{-}11)$$

Since the pressure outside such a bubble is not atmospheric, the symbol ΔP is used to represent the excess pressure inside the bubble.

The same basic equation would apply to a balloon but the γ would have to be replaced by an elastic tension T which is not constant and varies with the stretch of the material. However, at any equilibrium position of the balloon $\Delta P = 2T/r$ would apply if T is the elastic tension per unit length at that stretch.

The equation which we have derived for a sphere is a particular case of the more general equation for any curved surface with two radii of curvature:

$$\Delta P = \gamma\left(\frac{1}{r_1} + \frac{1}{r_2}\right) \qquad (9\text{-}12)$$

In the case of a sphere, both radii are the same and Equation (9-12) reduces to Equation (9-11).

EXAMPLE: SURFACE TENSION EFFECTS IN THE LUNGS

The most extensive surface of the body in contact with the environment is the moist interior surface of the lungs. To carry on the exchange of CO_2 and O_2 between circulating blood and the atmosphere in sufficient volume to sustain life requires one square metre of lung surface for each kg of body weight. For a normal adult this amounts to the area of a tennis court. The feat is accomplished by compartmentalizing the lungs into the tiny air sacs called alveoli.

The surface tension in the outermost single layer of molecules of the film of tissue fluid that moistens the surface of the lungs accounts for a large percentage of the elasticity of the alveoli which is necessary to breathing. This surface-active substance (surfactant) has a much lower surface tension than water so that it brings about a more even distribution of pressure between the large and small alveoli and reduces the overall pressure thereby decreasing the muscular effort required for respiration. Also, a lower surface tension permits a closer fit of the alveolar surface to the capillaries for maximum efficiency of transfer of oxygen, etc. Specifically, during inhalation the surface area of the lungs is enlarged and the surface concentration of lung surfactant decreases (i.e. the same number of molecules is spread more thinly) leading to an increase in the surface tension. During exhalation, the surface area of the lungs decreases leading to an increase in concentration of surfactant and thus a decrease in surface tension. This decrease in surface tension stabilizes the airways in the lungs preventing their collapse.

The surfactant in the lungs is a lipoprotein, that is, a compound molecule composed of protein and lipid constituents. In the lungs, the lipid has a polar head group which is hydrophilic and long fatty acid chains which are hydrophobic and are believed to stick up out of the surface.

In hyaline membrane disease prevalent in premature babies, the lipid is absent. The molecules of the liquid lining have a high surface tension and attract each other. The alveoli collapse after exhalation and must be re-expanded. Thus it becomes increasingly difficult for the infant to inflate its lungs. The baby dies as a result of oxygen starvation and exhaustion. Researchers at MIT have been investigating the use of a lipid aerosol which has proved fairly successful in getting these children over the first ten days or so until their body produces enough lipid.

9-4 Capillarity

If small droplets of different liquids are placed on different surfaces and examined carefully, a wide variety of shapes for the drops will be observed. For water on glass, the drop has the shape indicated in Figure 9-7a, while for mercury on glass the shape is as in Figure 9-7b. The different shapes arise because the liquid molecules have different affinities for the glass molecules. This behaviour is characterized by the contact angle θ which is shown in Figure

9-7. The contact angle is a measure of the curvature of the liquid-vapour interface at a solid surface. For water on clean glass $\theta \simeq 0°$ whereas for water on dirty glass $\theta \simeq 30°$ and for mercury on glass $\theta \simeq 140°$.

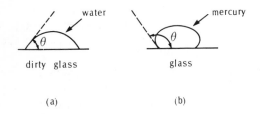

(a)   (b)

Figure 9.7
Definition of contact angle

Liquids having contact angles between 0 and 90° are said to wet the solid surface. Liquids having contact angles between 90 and 180° are said not to wet the solid surface.

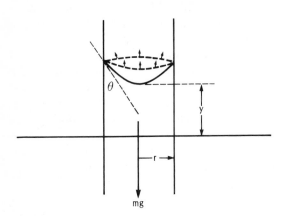

Figure 9.8
Capillary rise in a cylindrical tube

This wetting effect leads to the phenomenon of capillary rise. If tubes of small internal diameter are placed vertically in different liquids, the liquid will be observed to rise or fall in the tube depending on whether the contact angle is less than or greater than 90°. Con-

sider the column of water at equilibrium in the glass tube in Figure 9-8. The water has risen up the tube a distance y above the surface of the water in the container. There is force due to the surface tension which pulls up all the way around the edge of the meniscus. The vertical component of this force is

$$F = 2\pi r\gamma \cos\theta \qquad (9\text{-}13)$$

Gravity pulls down on the cylinder of liquid in the tube. Thus there is present a downward force, namely the weight of the column. This force is $F = mg = \text{volume} \times \rho \times g = \pi r^2 y\rho g$. When the liquid in the column is at rest, these two forces are equal. Then

$$2\pi r\gamma \cos\theta = \pi r^2 y\rho g$$

from which

$$y = \frac{2\gamma \cos\theta}{\rho g r} \qquad (9\text{-}14)$$

Note that if θ is greater than 90°, $\cos\theta$ will be negative. The negative y that results represents the depression of the liquid in the tube below the surface of the liquid in the container.

EXAMPLE: CAPILLARY RISE IN PLANTS

Capillarity plays an important role in the movement of water in plants. However, the vascular systems (xylem) of plants and trees are not open to the air at the upper end so capillary rise in the xylem of a plant is not exactly analogous to the situation just discussed. Nevertheless, taking the radius of the xylem as approximately 20 μm and a contact angle of zero, Equation (9-14) predicts that water could rise to a height of about 0.75 m. This is sufficient to get water to the top of small plants but not trees. However the walls of the xylem contain a network of passageways of quite small radius. These could support water columns of the order of km in height.

9-5 Viscosity

Fluids in motion experience an internal friction which tends to retard the motion of the fluid. This internal resistance is termed viscosity. Figure 9-9 shows two adjacent fluid layers each having a thickness dy. The upper layer has a speed which is dv greater than the lower. The frictional forces F between these two layers are shown. They are equal in magnitude but opposite in direction and parallel to the surface. Suppose this force F acts over an interlayer surface area A. Then the frictional or viscous force per unit area is F/A and is called the shear stress (Chapter 7) in the fluid.

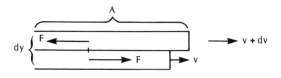

Figure 9.9

Frictional forces between two adjacent fluid layers

For many fluids, it is found that the shear stress at each point is proportional to the velocity gradient at that point, i.e.

$$F/A \propto dv/dy$$

or   (9-15)

$$F/A = \eta \, dv/dy$$

The proportionality constant, η, in Equation (9-15) is called the coefficient of viscosity or simply the viscosity of the fluid. Its value depends on the fluid and on such parameters as temperature and pressure. The dimensions of viscosity are $[\text{force}] \times [\text{time}]/[\text{length}]^2 = ML^{-1}T^{-1}$ and the units are $N\,m^{-2}\,s$. A more commonly used unit is the poise named in honour of Poiseuille, a nineteenth century French investigator of liquids. One poise $= 10^{-1}\,N\,m^{-2}\,s$. Some typical viscosities are given in Table 9-3.

It will be noted that the viscosity of liquids

Table 9-3
The Viscosity of Various Liquids

Fluid	Viscosity $(\eta)\,(N\,m^{-2}\,s)$
Water at 20°C	0.001 or 1 cP*
Water at 100°C	0.0003
Castor Oil at 20°C	1
Air at 20°C	18×10^{-6}
Air at 100°C	21×10^{-6}

* 1 cP = 1 centipoise = 10^{-2} poise.

decreases with increasing temperature, whereas that of gases increases. Viscosity in gases, where the molecules are far apart most of the time, results from diffusion of molecules between adjacent layers. Since the average molecular speed of random motion increases with temperature, diffusion and, hence, the viscosity of gases increases with increasing temperature (approximately as $\sqrt{T}$). For liquids where the molecules are close together at all times, viscosity arises from the work against intermolecular forces that each molecule must do to move ahead. Rising temperature increases the kinetic energy of random motion of the molecules and thus increases the relative number of molecules which have enough kinetic energy to overcome the intermolecular forces and this reduces the viscosity. Fluids which obey Equation (9-15) are called Newtonian fluids. These include gases and most common liquids—for example, water, alcohol, liquid metals.

There are many fluid and quasifluid systems which do not obey Equation (9-15). Thixotropic substances are fluid when they are in motion but solid when they are not, for example, wet sand. Dilatant substances are fluids for which the viscosity increases with increasing stress, for example suspensions of starch grains. From a biological point of view, one of the most important groups of non-Newtonian fluids are the viscoelastic liquids. For such substances the expression for the sheer stress is to a first approximation given by

$$F/A = M\varepsilon_s + \eta \frac{d\varepsilon_s}{dt} \qquad (9\text{-}15)$$

where M is the elastic bulk modulus of the liquid, ε_s is the strain and η is the viscosity of the liquid. In this case the velocity gradient is not proportional to the shearing stress. One of the most important of the viscoelastic liquids is synovial fluid which fills the cavities in the synovial joints, like the knee, of mammals. It is similar to blood plasma but has less protein and contains a long polysaccharide, hyaluronic acid. It is this acid which seems to control the viscosity of the liquid. The viscosity of synovial fluid decreases as the stress and velocity gradient increase which of course is necessary for smooth joint operation. Marked changes occur in the value of the viscosity of synovial fluid with the onset of certain diseases, in particular arthritis. Needless to say, there is at present a large research effort directed toward the investigation and characterization of such viscoelastic liquids.

PROBLEMS

1. At a certain height above sea level, the gravitational potential energy and the most probable kinetic energy (kT) of the air molecules are the same. What is the pressure at that height?
2. Calculate the height in the atmosphere where the atmospheric pressure is $\frac{1}{10}$ of that at sea level. Assume that the average molecular weight of the air molecules is 29 and that the air is at a uniform temperature of 300 K.
3. Calculate the pressure in atmospheres at the top of Mount Everest whose peak is 10 km above sea level. Assume that the ratio between the weight of a single air molecule and its thermal energy is

10^{-4} m^{-1}. Also assume sea level pressure to be 1.00 atm.
4. If ice has a density of 0.92×10^3 kg m^{-3} what is the average volume occupied by one molecule?
5. A biologist uses a chamber (C) to study the effect of air pressure on small organisms placed inside. The pressure in the chamber is measured with a mercury-filled U tube manometer (M) open at one end to the atmosphere as shown in the diagram. For the case illustrated, what is the absolute pressure in the chamber? (Take the density of mercury as 13.0×10^3 kg m^{-3} and atmospheric pressure as 10^5 Pa.)

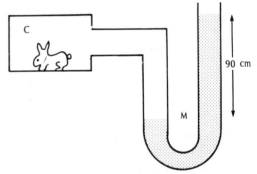

6. A person stands on his head. By how much has the hydrostatic pressure in his brain increased over the value when he is in the normal upright position? Use your own dimensions.
7. A typical human's systolic blood pressure is 120 mm of mercury. Express this in N m^{-2}. Is this an absolute or a gauge pressure?
8. With strong inspiratory effort, the gauge pressure in the lungs can be reduced to -80 mm Hg. Gin has a density of 0.22×10^3 kg m^{-3}. What is the greatest height that gin can be sucked up a straw?
9. What is the gauge pressure and the absolute pressure in chamber A? Assume both manometer liquids to be water.

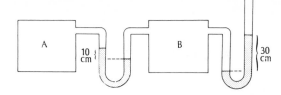

10. Calculate the gauge pressure inside a spherical bubble of radius 8×10^{-5} m located in a water tank 15 m below the surface of the water. On top of the water there is a 5 m layer of oil of density $0.5 \times 10^{+3}$ kg m^{-3}.

11. A bubble of gas located in water at a depth of 10 metres has a volume of 4×10^{-9} m^3. A smaller bubble of the same gas located at the same depth has a volume of 0.5×10^{-9} m^3. The pressure in both bubbles exceeds the pressure of the water outside the bubbles at that depth. What is the excess pressure of the smaller bubble, compared with the excess pressure of the larger bubble?

12. A biologist has a sample of lung surfactant and wishes to measure its surface tension. He first determines that its density is 80 percent of the density of water and that its contact angle with glass is the same as that for pure water. He places a capillary tube in a sample of water and notes that the capillary rise is 0.10 m. He then places the same tube in surfactant and finds the rise to be only 0.02 m. What is the surface tension of the surfactant if the surface tension of pure water is 70×10^{-3} N m^{-1}?

13. A capillary tube is dipped in water with its lower end 0.05 m below the water surface. Water rises in the tube to a height of 0.02 m above that of the surrounding liquid and the contact angle is zero. What gauge pressure is required to blow a hemispherical bubble at the lower end of the tube?

14. A liquid rises in a capillary as shown in the figure. Its density is 8×10^3 kg m^{-3}. What is (a) the gauge pressure at A, (b) the gauge pressure at B?

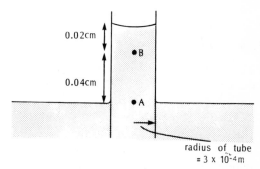

radius of tube $= 3 \times 10^{-4}$ m

15. A capillary tube is placed in water and the water is observed to rise up in the tube to a height of 10 cm. The tube is then lowered further into the water so that only 8 cm of tube is above the water surface. Will the water:
 (a) jet up like a fountain to a height of 2 cm above the top of the tube?
 (b) flow over the top of the tube and run down the sides?
 (c) rise to the top and stop?

16. A thin film of oil is spread over a very flat floor. On the oil floats a large flat plate of area 4 m^2. Suppose the oil's thickness is 100 μm. What force would have to be applied to the floating plate to move it at a speed of 0.01 m s^{-1}?

Chapter 10

Motion of Molecules
in Solution

10-1 Introduction

About 1827, Robert Brown first observed the random chaotic motion of pollen grains in solution. This was the first visible evidence of the kinetic theory at work in liquids. The particles would sometimes change speed and direction without apparently colliding with anything. Brown realized that collisions between the pollen grains and water molecules, which have their own similar but invisible chaotic motion, were influencing the motion of the pollen grains.

Any molecule in solution, whether invisible or not, will undergo this Brownian motion. If we could only watch and measure the changes in velocity of one molecule then we would be able to learn a great deal about the molecule. For instance, if the velocity was always small then conservation of momentum laws would tell us that we were dealing with a large molecule. Similar momentum exchanges with smaller molecules would leave them travelling with higher velocities. Thus the degree of motion or diffusion of particles in a solvent decreases with increasing particle size. This property of diffusion is vital to the orderly functioning of biological cells. In fact more than any other process, diffusion has determined basic properties such as the ultimate size of the cells.

In principle, there are two ways we could investigate diffusion. If we could pick out one particle or molecule and could follow its motion for a very long time we could learn about all of its diffusive properties. Alternatively we could investigate the same properties by viewing a large number of identical particles. These two methods, known as the time approach and the ensemble approach, lead to the same average results. We will use the latter to discuss several of the dynamic processes related to diffusion.

10-2 The Random Walk

The diffusion of molecules is basically a random walk process. To illustrate the phenomenon in simple terms consider the motion of a bunch of drunks ejected from a bar. The common approach would be to follow the position of a drunk with time. We'll take the ensemble approach—that is, consider the whole bunch of drunks. Let's assume that they are all totally inebriated, but able to stumble around when the bar closes. They are all forcibly ejected onto the street as in Figure 10-1 at time $t = 0$.

Figure 10.1
Drunks ejected from bar

With time these drunks begin to move up and down the street, but in a completely random way. Some may start up the street, stumble around and then go down the street. Eventually (unless another bar opens) they will be equally distributed up and down the street. The position of the drunks, with time, is shown in Figure 10-2.

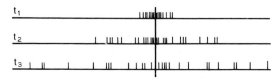

Figure 10.2

Distribution of random walking drunks with time

Note that the average displacement ($\bar{x}$) of the drunks is zero because as many go up the street as go down. Thus it is useless for us to talk of the average or mean displacement of the drunks because this value is always zero. Instead, it is convenient to consider the root mean square displacement ($\sqrt{\overline{x^2}}$), a quantity which is always positive and increases with time. The behaviour of particles in solution can be analysed in the same fashion. The root mean square displacement for n identical particles is

$$x_{\text{rms}} = \sqrt{\overline{x^2}} = \left(\frac{\sum x^2}{n}\right)^{1/2} \quad (10\text{-}1)$$

After many steps are taken by many particles, the distribution of particles relative to their initial starting point or points becomes a Gaussian distribution. This is shown in Figure 10-3.

In this graph, each vertical line corresponds to the number (N) of particles found between x and $x + \Delta x$, Δx being the distance between the lines. Alternatively, the smooth line joining the tops of the bars may be considered as a representation proportional to the density of particles along x. If the curve is drawn carefully and the areas inside and outside the limits of $\pm\sqrt{\overline{x^2}}$ measured, it is found that approximately $\frac{2}{3}$ of the particles travel less than this distance and $\frac{1}{3}$ travel more. We will see why the function has a Gaussian shape in the next section. In a three dimensional case real particles have a mean square displacement of $\overline{R^2}$ where

$$\overline{R^2} = \overline{x^2} + \overline{y^2} + \overline{z^2} \quad (10\text{-}2)$$

Thus, while it is completely impossible to predict what one drunk or particle will do, we can predict what will happen to the distribution of the group.

10-3 The Role of Thermal Forces

One major flaw in the drunks analogy of the previous section is that they were totally "self-propelled." In the real situation in solution, particles, unless they are motile swimmers like spermatozoa, move due to thermal forces in their environment. To improve our analogy, we may consider what might have happened if the two theaters on either side of the bar closed just before the bar and the street was teeming with children. Effectively these running, scrambling children are the solvent for the larger drunks. We can imagine that the energy of these little monsters is so great that they buffet and carry the drunks all up and down the street. We note that motion of the

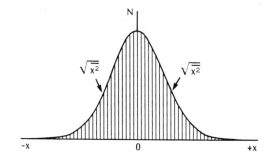

Figure 10.3

Gaussian distribution of particles

heavier drunks would not be perturbed as much as the ligher ones. Even if a big drunk was given a tremendous push he would quickly suffer more retarding collisions than a small one. If all the drunks were the same size, however, they would end up being distributed as in Figure 10-3. The time it took to reach a distribution such as shown in Figure 10-3 would depend upon the energy of the children and drunks as they bounced and scrambled about.

In gases and in solutions, the amount of energy involved in the dynamic processes of translation and rotation is determined by temperature. However, the energy is not equally distributed over all the particles present. Some particles will have considerably more than their fair share of the energy and some will have less. Their average energy does depend on temperature. This distribution of the energy was first demonstrated in 1890 by the French chemist Jean Perrin. He reasoned that Brown's dancing pollen grains (or anything similar) should move about and be governed by the same laws as the molecules in a gas. He argued that, just as the number of molecules in the atmosphere decreases exponentially with elevation (as evidenced by changes in pressure, see Chapter 9), then the number of particles in solution should decrease exponentially from bottom to top. He set out to prove experimentally that this decrease existed. Obviously his particles, whatever he used, had to be denser than water or they would all float to the top. If they were exactly the same density as water then, by Archimedes' principle, they would be apparently weightless and stay wherever they were placed. If denser than water, they had to be large enough to be seen but small enough that collisions with water molecules could provide sufficient energy to keep them all from falling to the bottom. He was able to achieve this requirement by making suspensions of gam-

boge (tree resin) particles of uniform size (diameters between 10^{-6} and 10^{-7} m) and observing them with a microscope. The experiment was carried out at 15°C (288 K) with gamboge particles of average radius 0.22×10^{-6} m and density $\rho = 1.20 \times 10^3$ kg m^{-3} suspended in water of density $\rho_\ell = 1.00 \times 10^3$ kg m^{-3}.

Figure 10-4 illustrates the relative numbers of particles counted at successive 30 μm intervals from the bottom up after the suspension of the particles had been well shaken and allowed to reach an equilibrium distribution.

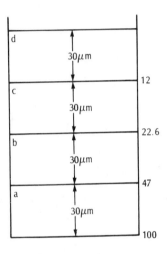

Figure 10.4

Gamboge particles in a suspension

If the logarithms of the relative numbers of particles are plotted against the distance from the bottom, a straight line will be obtained indicating that there is an exponential decay analogous to that of atmospheric pressure.

By comparison with Equation (9-7), we can write expressions for the relative number of particles in each box of Figure 10-4 as follows

$$N_d/N_a = e - (E_d - E_a)/kT \qquad (10\text{-}3)$$

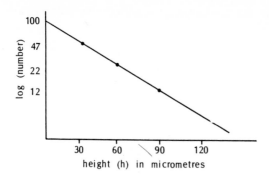

Figure 10.5

Semi-log plot of data in figure 10.4

or

$$\ln N_d/N_a = \frac{-(E_d - E_a)}{kT} \qquad (10\text{-}4)$$

In these equations, N_a corresponds to the number of particles in box a, each having an energy, E_a. N_d and E_d are similarly defined. Note that we can write

$$\Delta E = E_d - E_a$$

so that Equation (10-4) becomes

$$\ln N_d/N_a = -\Delta E/kT \qquad (10\text{-}5)$$

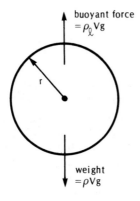

Figure 10.6

Buoyancy effect in solution

This extremely well-known relation is called the Boltzmann equation. It completely describes the way energy is distributed among identical particles. By evaluating ΔE we can use Perrin's data of Figure 10-4 to determine Boltzmann's constant, k.

If V is the volume of each particle then the net downward force acting on it will be (see Figure 10-6)

$$F = mg = gV(\rho - \rho_\ell)$$

The potential energy of the particles is then

$$E = Fh = (\rho - \rho_\ell)Vgh$$

where h is the height of the particles from the bottom of the container. For the situation represented in Figure 10-4, the Boltzmann equation becomes

$$\ln \frac{N_d}{N_a} = \frac{-[Vgh_d(\rho - \rho_\ell) - Vg(0)(\rho - \rho_\ell)]}{kT}$$

We know the following quantities

$$h_d = 90 \times 10^{-6}\ \text{m}$$
$$T = 288\ \text{K}$$
$$N_d = 12$$
$$N_a = 100$$
$$V = \tfrac{4}{3}\pi r^3 = (\tfrac{4}{3})\pi(0.22 \times 10^{-6})^3$$
$$= 4.5 \times 10^{-20}\ \text{m}^3$$

Thus

$$k = \frac{-((\rho - \rho_\ell)\,Vgh_d)}{T \ln (N_d/N_a)}$$

$$= \frac{-(1.20-1.00)\times 10^3 \times 4.5 \times 10^{-20} \times 9.8 \times 90 \times 10^{-6}}{288\ \ln (12/100)}$$

$$= 1.3 \times 10^{-23}\ \text{J K}^{-1}$$

This value for the Boltzmann constant is quite close to the presently accepted value of $k = 1.38 \times 10^{-23}\ \text{J K}^{-1}$. If we had considered many experiments and had made good use of the data by finding the slope of the best-fit line as in Figure 10-5, our agreement would have been even better.

EXAMPLE

Using an experimental setup similar to that shown in Figure 10-4, the number of particles in the lowest box was 90 at 300 K. How many particles would you expect to find in box c? (Use $k = 1.38 \times 10^{-23}$ J K^{-1})

$$\ln \frac{N_c}{N_a} = -\frac{Vgh_c(\rho - \rho_\ell)}{kT}$$

$$\ln \frac{N_c}{90} =$$

$$-\frac{4.5 \times 10^{-20} \times 9.8 \times 60 \times 10^{-6} (1200 - 1000)}{1.38 \times 10^{-23} \times 300}$$

$$N_c = 25$$

The previous examples have shown how potential energy is distributed among particles. The same theory applies to the distribution of kinetic energy. Some particles may obtain energy from the medium and increase their energy, but other particles will lose energy. Thus energy is continually exchanged between particles and the medium, but the distribution of energy as given by the Boltzmann equation remains essentially constant.

The Boltzmann equation, when applied to the kinetic energy of particles, can tell us quite a bit about the distribution of particle velocities and about diffusion. In this case

$$\ln \frac{N_x}{N_o} = \frac{-(E_x - E_o)}{kT} \quad (10\text{-}6)$$

where N_x and N_o are the numbers of particles having kinetic energies E_x and E_o respectively. Since kinetic energy has the form $mv^2/2$ then Equation (10-6) becomes

$$\ln \frac{N_x}{N_o} = \frac{-(mv_x^2 - mv_o^2)}{2kT}$$

If the lower energy state corresponds to $v_o = 0$ then

$$\ln \frac{N_x}{N_o} = \frac{-mv_x^2}{2kT} \quad (10\text{-}7)$$

This function is called a Gaussian function and has the shape shown in Figure 10-7.

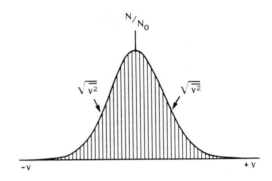

Figure 10.7
Gaussian distribution of velocities

In this diagram, the vertical lines indicate the fraction of particles having a velocity between v and $v + \Delta v$, where Δv is the separation of the lines.

If the function looks familiar, it has exactly the same shape as the function which described the position of the drunks on the street a short while after they were ejected from the bar (Figure 10-3). This, of course, makes sense because the displacement of the drunks or particles is linearly related to their velocity. Thus, we see that according to the Boltzmann equation, diffusing particles in solution have Gaussian-shaped distributions of velocities and displacements.

When dealing with average particle velocities in solution we have exactly the same problem as we had with displacement because the average velocity, $\bar{v}$, is zero. This is because as many particles have a negative velocity as have a positive velocity. Thus it is convenient to use root mean square velocities, $\sqrt{\overline{v^2}}$ instead of average velocities. This root mean square velocity has been shown in Figure 10-7. Note that, in comparison to the displacement case, one third of the particles are travelling faster than this velocity and two thirds slower. Again, we cannot predict the velocity of a single particle. All we know is the distribution of velocities of many particles.

Statistical physics texts such as those written by F. Reif show that when one considers the stepwise motion of the diffusers as they move from collision to collision the relationship between mean square displacement, $\overline{R^2}$, the mean square step length, $\overline{L^2}$, and the mean square velocity, $\overline{v^2}$, is given by

$$\overline{R^2} = (N\overline{L^2})t = \frac{6(\overline{v^2}\tau)t}{3} \qquad (10\text{-}8)$$

In this expression, t is time and N is the average number of steps taken per unit time. The symbol τ represents the time taken for each step or, in other words, the time between collisions.

It should be quite obvious from all this that the environment of biological macromolecules in a living cell is an extremely shaky one. If water molecules can keep the very large (in comparison to proteins) gamboge particles of Perrin's experiment in suspension then think about how proteins, ribosomes, etc. must be buffeted about. Imagine, if you will, an assembly line working in the midst of an earthquake and you will have some small idea of the problems of such well-adjusted systems as the assembly of proteins at ribosomes in the midst of random thermal bombardment by water (and other) molecules.

10-4 The Diffusion Coefficient

The quantity $(\overline{v^2}\tau)/3$ is often called the diffusion coefficient, D. Consequently Equation (10-8) can be written;

$$\overline{R^2} = 6Dt \qquad (10\text{-}9)$$

Equation (10-9) is very useful especially when one is trying to determine whether or not many of the processes of the biological cell, such as transcription and translation make sense in physical terms. For example one might ask whether or not the rates of diffusion are sufficient to allow 50 amino acids per second to be made into protein at the ribosome. As we'll note from the following two examples, the times which small particles require to diffuse across a cell are very small.

EXAMPLE

How long will it take a water molecule to diffuse 0.01 m through water? This question is typical of the rather imprecise language that is frequently encountered in this area. The question should really be interpreted as: how long will it take for one third of the water molecules in a considerable population to diffuse at least 0.01 m from their initial starting points?

The diffusion coefficient of water molecules in water near room temperature is $D \approx 2 \times 10^{-9}$ m^2 s^{-1}. Since

$$\overline{R^2} = 6Dt$$

then

$$t = \frac{\overline{R^2}}{6D} = \frac{(0.01)^2}{6 \times 2 \times 10^{-9}}$$

$$= 8,300 \text{ s}$$

$$= 2.3 \text{ hours}$$

This makes diffusion seem a very slow process indeed. However, it should be remembered that within the confines of a living cell the distances to be travelled are normally very small. Consider now the same question changing the distance from 0.01 m to the length of a bacterial cell (i.e. 3 μm). Then

$$t = \frac{\overline{R^2}}{6D} = \frac{(3 \times 10^{-6})^2}{6 \times 2 \times 10^{-9}}$$

$$= 7.5 \times 10^{-5} \text{ s}$$

The process in a real bacterial cell would not be quite this rapid since the cytoplasm will be about five times as viscous as water. This will decrease the diffusion constant to one fifth of the value used, with the result that the time will be increased by a factor of five to 3.8×10^{-3} s which is still very fast. Thus diffusion, while a slow process in a bulk liquid, is a very fast process within the confines of a cell.

EXAMPLE

Carbon monoxide hemoglobin (molecular weight 68,000) has a diffusion constant of 6.2×10^{-11} m^2 s^{-1} in water. In more viscous cytoplasm D would be $(6.2 \times 10^{-11})/5$ m^2 s^{-1} and the time to "travel" the 3 μm length of a bacterial cell will be

$$t = \frac{\overline{R^2}}{6D} = \frac{5(3 \times 10^{-6})^2}{6 \times 6.2 \times 10^{-11}}$$

$$= 0.12 \text{ s}$$

The rate at which a particle diffuses has just been shown to be strongly dependent on the parameter D which has been called the diffusion coefficient. The actual significance of this parameter is better treated by a consideration of Fick's law

$$J = -D \frac{dc}{dx} \qquad (10\text{-}10)$$

This equation states that J, the flux of particles (number of particles passing through an imaginary normal surface of unit area) is related to the force which is pushing them, $(-dc/dx)$. In simplest terms, this driving force is just a gradient in concentration causing particles to move from regions of high concentrations to regions of low concentration. The steeper the gradient (the bigger the concentration difference) the greater will be the flux. The diffusion coefficient D is the proportionality constant in Fick's law. It provides a measure of both the frictional resistance experienced by the particle as it moves through the solvent and its kinetic energy due to thermal effects (Equation (10-7)).

The Stokes-Einstein equation for D is

$$D = \frac{kT}{f} \qquad (10\text{-}11)$$

where f is the friction factor of the particle. For a spherically shaped particle

$$f = 6\pi\eta r \qquad (10\text{-}12)$$

where η is the coefficient of viscosity of the solvent and r is the radius of the particle.

The diffusion coefficient is essentially independent of concentration although at higher concentrations it may be influenced by intermolecular interactions. Thus D is the same if particles are uniformly spread through the solution or if they are more localized, for example, on one side of a membrane. In the first case the concentration gradients (dc/dx) are fluctuating and microscopic. In the latter case they are macroscopic. Because of this the latter case is better for simple measurements of D.

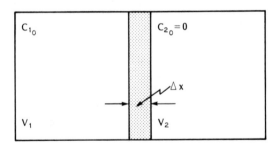

Figure 10.8
Two compartment container for diffusion experiments

Suppose two liquid compartments of volumes V_1 and V_2 are separated by a porous membrane of thickness Δx. The volume V_1 contains a solution of the material whose diffusion constant we wish to determine. The initial concentration in this side will be designated C_{1_o}. The other volume, V_2, initially contains pure water and thus its initial concentration will be $C_{2_o} = 0$.

If the area of the membrane is A, then the flux of particles through the membrane is given by

$$J = \frac{1}{A}\left(\frac{dc}{dt}\right) \qquad (10\text{-}13)$$

Then application of Fick's law leads to

$$\frac{dc}{dt} = -DA\frac{dc}{dx} \qquad (10\text{-}14)$$

We can approximate the concentration gradient (dc/dx) across the membrane by $(C_1 - C_2)/\Delta x$ and then

$$\frac{dc}{dt} = \frac{-DA}{\Delta x}(C_1 - C_2) \quad \text{for} \quad C_1 > C_2 \quad (10\text{-}15)$$

This equation can be solved to obtain an equation which describes the concentration, C_1, in V_1 as a function of time. If the two volumes are equal this is:

$$C_1 = \frac{C_{1_o}}{2}e^{-2DAt/\Delta x V_1} + \frac{C_{1_o}}{2} \qquad (10\text{-}16)$$

Every variable except D is directly measureable and substitution yields a value of the diffusion constant across the membrane. Usually, however, what is required is not D across the membrane but D in pure solution. This can be determined by correcting the experimentally measured D for the percentage of the membrane area that actually consists of pores. Thus, if the area were 50 percent pores then the true diffusion constant would be twice that experimentally determined and if the area were only 5 percent pores then the measured D would be increased by 20 times. This type of correction will be fairly valid if the pores are straight and have a diameter significantly larger than that of the diffusing molecules. In addition, surface changes on the

pores and diffusing molecules should be small or non-existent.

Clearly it would be preferable to have a system which measures the diffusion constant directly. This can be done in a system where monochromatic laser light is scattered from a solution of the molecules in question. Since the scattering molecules are undergoing Brownian motion, some of the incident photons will scatter from molecules approaching the photons at a variety of velocities. As a result, the frequency of many incident photons is Doppler shifted to higher and lower values so that the incident monochromatic light emerges as scattered light with a definite distribution about the incident frequency. The width of this distribution is directly related to the diffusion constant of the solute being investigated. This is probably the best method available, but requires sophisticated instrumentation not available in most laboratories.

If a diffusion constant has been measured and we assume the molecule to be a sphere then since $D = kT/f$ and $f = 6\pi\eta r$ (assumed), the radius of the molecule will be:

$$r = \frac{kT}{6\pi\eta D} \qquad (10\text{-}17)$$

The volume of a sphere is $V = \frac{4}{3}\pi r^3$ or

$$V = \frac{4}{3}\pi\left(\frac{kT}{6\eta\pi D}\right)^3 \qquad (10\text{-}18)$$

and the mass of a molecule will be $m = \rho V$ where ρ is the density of the molecule. Finally, the molecular weight will be $M = mN_A$. Combining the foregoing equations we obtain:

$$M = \rho\frac{4}{3}\pi\left(\frac{kT}{6\pi\eta D}\right)^3 N_A \qquad (10\text{-}19)$$

Molecular weights calculated by this method will frequently be too large because, as a result of the presence of charged groups on their surfaces, most biological molecules will pick up a monolayer of water molecules equivalent to about 0.35 g of water per gram of protein. Even after corrections are made for bound water, an error will usually remain which can indicate how non-spherical the molecule may be. For example, hemoglobin has an accepted molecular weight of 68,000 but the previous equation gives 130,000. This is reduced to 89,000 by a correction for bound water yielding a result slightly high, indicating that the molecule is slightly non-spherical. At the other extreme tobacco mosaic virus (TMV) is a long thin rod of length 280 nm and diameter 15 nm. Its accepted molecular weight is 40×10^6 but based on diffusion constant measurements a value of 230×10^6 is obtained which is reduced to 160×10^6 upon correction for bound water. This very large discrepancy immediately tells us that the molecule is very non-spherical. Thus diffusion measurements can be a powerful source of information on various physical properties of macromolecules.

10-5 Sedimentation

If a sample of blood is removed from some animal and allowed to stand after addition of a suitable anticoagulant such as heparin, it will be observed that a clear slightly yellowish zone will appear and grow at the top of the container. Microscopic examinations would reveal that the erythrocytes (red blood cells) are slowly settling out because of the earth's gravitational force on them. Within quite wide ranges for normal individuals the settling rate for an average human male will be about 3 mm hr^{-1} and for a female 4 mm hr^{-1}. In some disease conditions, this can increase drastically to around 100 mm hr^{-1}. Different animal species often show marked differences in settling rates. For samples from horses the rate is rapid while for samples from cows the rate is slow.

The erythrocyte is normally a biconcave

disk as illustrated in Figure 10-9, but for approximate calculations it can be assumed to be a sphere of radius 3 μm. As it falls through the plasma it experiences fluid friction. This fluid frictional force is proportional to the settling velocity, v. The constant of proportionality is the frictional coefficient, f, which has been shown to have the form $6\pi\eta r$ for spherical particles. Thus

$$F_v = fv = 6\pi\eta rv \qquad (10\text{-}20)$$

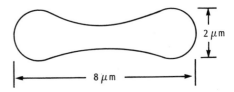

Figure 10.9

Dimensions of a human erythrocyte

Gravitational and buoyant forces will also act upon the erythrocyte. The gravitational (downward) force will be

$$F_g = mg = \rho_s Vg = \rho_s \tfrac{4}{3}\pi r^3 g \qquad (10\text{-}21)$$

where ρ_s is the density of the settling object and V is its volume. The buoyant (upward) force can be calculated using Archimedes' principle as:

$$F_b = \rho_l \tfrac{4}{3}\pi r^3 g \qquad (10\text{-}22)$$

where ρ_l is the density of the displaced fluid.

At terminal velocity, v_t, the upward and downward forces are equal. Thus

$$F_b + F_v = F_g \qquad (10\text{-}23)$$

or

$$\rho_l \tfrac{4}{3}\pi r^3 g + 6\pi\eta r v_t = \rho_s \tfrac{4}{3}\pi r^3 g \qquad (10\text{-}24)$$

from which

$$v_t = \frac{2}{9}\frac{r^2 g}{\eta}(\rho_s - \rho_l) \qquad (10\text{-}25)$$

EXAMPLE

What happens to the blood cells in a disease where the sedimentation rate suddenly increases from 4 to 100 mm hr^{-1}?

Obviously the densities cannot change significantly and direct experimentation shows negligible change of viscosity. Therefore

$$\frac{v_t(\text{sick})}{v_t(\text{healthy})} = \frac{r^2(\text{sick})}{r^2(\text{healthy})} = \frac{25}{1}$$

from which r (sick)$/r$(healthy) $= 5/1$.

The erythrocyte membrane is relatively inextensible so the only way the apparent radius can increase by a factor of five is for the cells to gather in groups of some sort which can then settle out faster. Microscopic examination of the blood from the diseased patients reveals that the erythrocytes stack up in arrangements called rouleaux which resemble stacks of coins.

The sedimentation of erythrocytes, which we have just discussed, is successful because the particles are fairly large. This being the case, the gravitational force has more influence on their motion than the diffusion type forces. If one wants to study sedimentation of smaller particles, it is necessary to increase the settling force substantially. This can be done by the use of a centrifuge. The sample is spun rapidly at a radius, R, from the axis of the rotor. Once the rotor is spinning, the centrifugal forces are much greater than gravitational forces and the particles tend to settle outward. Once terminal velocity is reached, the outward and inward forces must balance as they did in the erythrocyte case. Thus

$$F_v + F_b = F_c \qquad (10\text{-}26)$$

where F_c is the centrifugal force. If the rotor is turning at a frequency of ν cycles per second, then the centrifugal force on a particle of mass m is

$$F_c = m4\pi^2\nu^2 R \qquad (10\text{-}27)$$

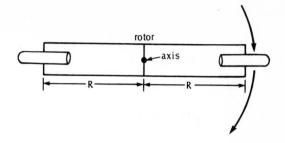

Figure 10.10

Top view of the rotor of a centrifuge

The buoyant force, F_b, is equal to the force on the mass of solution (m_o) which is displaced by the particle or

$$F_b = m_o(4\pi^2\nu^2)R \qquad (10\text{-}28)$$

At terminal velocity v_t, the retarding or frictional force F_v is as before

$$F_v = fv_t \qquad (10\text{-}29)$$

where f is the friction coefficient defined earlier. Thus Equation (10-26) becomes

$$fv_t + m_o(4\pi^2\nu^2)R = m(4\pi^2\nu^2)R \qquad (10\text{-}30)$$

or

$$v_t = \frac{4\pi^2\nu^2 R}{f}(m - m_o) \qquad (10\text{-}31)$$

The mass of the particle is given by Equation (10-21) and the mass of the solution displaced can be approximated for dilute solutions, by

$$m_o = \rho_l[(\tfrac{4}{3})\pi r^3]$$

where ρ_l is the density of the solvent. If we also assume that the particle is spherical so that its friction coefficient is that of Equation (10-12), then

$$v_t = \frac{8(\pi^2\nu^2 R)r^2(\rho_s - \rho_l)}{9\eta} \qquad (10\text{-}32)$$

This equation for terminal velocity in a cen-

Table 10-1
Some Sedimentation Coefficients

Substance	s (seconds)	Mol. Wt.
Lysozyme	1.91×10^{-13}	14,400
Bovine Serum Albumin	5.01×10^{-13}	66,500
Bushy Stunt Virus	132×10^{-13}	10,700,000
Tobacco Mosaic Virus	193×10^{-13}	32,000,000

trifuge should be compared to Equation (10-25), which describes the terminal velocity of a particle settling due to the gravitational force.

A more commonly used parameter is the sedimentation coefficient, s. This, for the centrifuge case is

$$s = \frac{v_t}{4\pi^2 \nu^2 R} \qquad (10\text{-}33)$$

The quantity s is then a characteristic of the particle only and is independent of the rotor frequency or the centrifuge used. Combining Equations (10-32) and (10-33) we obtain

$$s = \frac{2}{9} \frac{r^2}{\eta} (\rho_s - \rho_l) \qquad (10\text{-}34)$$

Sedimentation coefficients for a few well known biological particles are given in Table 10-1.

Problems

1. Suppose that the position of each molecule in a system of molecules could be observed as a function of time. At a certain time t, the displacement of each molecule relative to its displacement at $t = 0$ is determined. The table below gives data for the number of molecules found as a function of the x-component of the displacement.

   For the x-component of the displacement, calculate

# of molecules (thousands)	x-component of displacement (m)
2	-0.045 to -0.035
4	-0.035 to -0.025
7	-0.025 to -0.015
9	-0.015 to -0.005
10	-0.005 to $+0.005$
9	0.005 to 0.015
7	0.015 to 0.025
4	0.025 to 0.035
2	0.035 to 0.045

   (a) the average value
   (b) the mean square value
   (c) the rms value
   (d) the most probable value

   Sketch a histogram showing the distribution of molecules as a function of the x-component of displacement.

2. In an experiment similar to Perrin's, 100 particles per unit volume were counted close to the bottom of a column, and 1 cm above this, 50 particles were counted in a similar volume. How many particles per unit volume would be expected at a height of 5 cm above the bottom?

3. In an experiment similar to Perrin's, the number density of particles at height 10 cm compared to the number density at the bottom is $10^{-4} : 1$. What is the ratio of the number density at height 5 cm to the number density at the bottom?

4. The diffusion constant of water in water is 2×10^{-9} m^2 s^{-1}. How long will it take for $\frac{1}{3}$ of the water molecules in a large popu-

lation to diffuse at least 1 mm from their initial position?

5. A biological macromolecule has a diffusion constant of 1.5×10^{-11} m² s⁻¹. If the position of each molecule in a solution were known at some instant then 1 s later how far will approximately $\frac{1}{3}$ of the molecules have travelled?

6. A spherical molecule has a diffusion constant of 5.0×10^{-11} m² s⁻¹. How long will it take $\frac{1}{3}$ of such molecules to diffuse a distance of 1 micrometre (10^{-6} m) in an E coli cell where the viscosity is 5 times as great as the viscosity of water?

7. Consider two different protein molecules which are both spherical but one has a molecular weight which is 8 times the other. What is the diffusion coefficient in water of the larger molecule relative to the smaller?

8. Molecules can diffuse across living cells in very small fractions of a second. If the width of average cells were increased to 10 times their average width then the average time required for molecules to diffuse across them would change by what factor?

9. If the diffusion coefficient for CO-hemoglobin in water is 6.2×10^{-11} m² s⁻¹ and for CO-hemoglobin in an unknown liquid is 15.5×10^{-11} m² s⁻¹, what is the viscosity of the liquid?

10. The diffusion constant of a large protein molecule is found to be 2×10^{-13} m² s⁻¹ in water at 20°C. Assuming it is spherical with a density of 1.3×10^3 kg m⁻³ compute its molecular weight.

11. A certain kind of spherical molecule is observed to diffuse through water so that $\frac{1}{3}$ of the molecules cover at least 0.5 cm in about 10 hours at 20°C. Compute the effective radius of the molecule.

12. A bacterium has a mass of 2.0×10^{-15} kg and a density of 1.05×10^3 kg m⁻³. When submerged in water (density 1.00×10^3 kg

m⁻³), what would be its apparent weight?

13. A block of ice is held 2 metres below the surface of the water. It is then released. What is its initial acceleration? If its mass is 2 kg and if the Stokes-Einstein frictional factor is 3 kg s⁻¹ what is the terminal upward velocity? (Use density of ice = 920 kg m⁻³.)

14. Sand is defined as mineral particles of diameter 0.05×10^{-3} m to 2.0×10^{-3} m; clay is defined as mineral particles of diameter less than 2.0×10^{-6} m.

   (a) What is the ratio of the terminal velocity of the largest sand and largest clay particles, sedimenting in water? Assume the particles are spherical and have identical densities and that Stokes-Einstein equation is valid.

   (b) How long will it take the largest clay particles assumed spherical to settle 0.01 m in water? The density of clay minerals is about 2.6×10^3 kg m⁻³.

15. A centrifuge is a device which is used to speed up the sedimentation of particles by increasing the effective "g" factor. How much faster will a particle sediment if the "g" factor is increased by 10 times.

16. Consideration of the gravitational, buoyant and viscous forces on a sphere falling in a fluid leads to this expression for terminal velocity:

$$v_t = \frac{2}{9} \frac{r^2 g}{\eta} (\rho_s - \rho_l)$$

   Two spheres of the same size but different material have terminal velocities in the same medium in the ratio of 9 to 1. If the slower has a density twice that of water, what is the density of the faster sphere? (Assume density of water is 1.0×10^3 kg m⁻³.)

17. What is the effect of: (a) water of hydration, (b) non-spherical shape on the diffusion constant of a molecule, and on the molecular weight calculated from the diffusion constant?

Chapter 11
Fluids in Motion

11-1 Introduction

The biophysicist is often interested in fluids in motion or in objects moving through fluids. For example the process of breathing, the circulation of the blood, cytoplasmic streaming, the movement of living creatures through air or water, all involve fluid flow.

11-2 Streamline Flow

Many of the flow situations met in biological systems are of a fairly simple type known as streamline flow (sometimes called laminar flow). Streamline flow can be steady as in the flow of sap through the xylem of a tree or pulsed as in blood flow. The discussion in this section will be restricted to steady streamline flow. Pulsatile flow will be discussed in a later section.

The simplest possible case is shown in Figure 11-1(a). A fluid is flowing through the tube at a low velocity. It is assumed that the fluid, of density ρ kg m^{-3}, is incompressible, that there is no gain or loss of heat energy and that diffusion of molecules in the moving fluid has a negligible effect. If a small quantity of ink were released from a hypodermic needle inserted at A, ink would be observed to be swept downstream in a "streamline" parallel to the sides of the tube. If the process were repeated at B there would be a similar streamline but careful observation would reveal that this central streamline was travelling

faster. Obviously some velocity distribution exists across the opening of the tube. With great care a few molecules of the ink might be released right at the wall and a few of them would be seen to adhere to the wall, indicating that velocity decreases to zero at the wall.

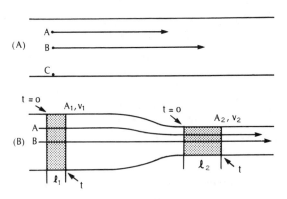

Figure 11.1
Streamline flow in a tube

Next the process will be repeated with a tube that narrows down, Figure 11-1(b). Similar streamlines will form but they will be forced to converge as the tube constricts. Again careful observation reveals that all streamlines speed up as the tube converges but that there is still a velocity distribution across the tube with central streamlines travelling faster than those near the walls. An expression describing the velocity distribution is required and will shortly be developed. However, it is frequently convenient to speak of "the velocity in a tube." It should be understood when this is done that it is really the average velocity that is being referred to.

If a pipe is rigid, then, regardless of its shape or change of shape, the amount of (incompressible) fluid that enters one end will be the amount that exits from the other end. This observation leads to a very useful conclusion. Consider Figure 11-1(b). The same volume of fluid must pass points in the wide and narrow portions of the tube in the same time interval t. Remember, the fluid is assumed incompressible. This volume is given by $A_1\ell_1 = A_2\ell_2$. The distance $\ell_1 = v_1 t$ and $\ell_2 = v_2 t$. Consequently:

$$A_1 v_1 t = A_2 v_2 t$$

or

$$A_1 v_1 = A_2 v_2 = \text{a constant} \qquad (11\text{-}1)$$

This very useful equation is known as "the equation of continuity." Note that Av has dimensions of volume per unit time and can be represented as the flow rate (Q):

$$Q = Av \qquad (11\text{-}2)$$

Sometimes in flow systems situations are encountered where a single "pipe" can divide (bifurcate) into 2 or more "pipes" as when an artery branches. The reverse can also occur as when venules join to form veins. In all of these situations the total flow in must equal the total flow out. Thus if n_1 tubes each of area A_1 with flow v_1 joined and then divided into n_2 tubes each of equal area A_2 with flow v_2 the equation of continuity could be modified as:

$$n_1 A_1 v_1 = n_2 A_2 v_2 \qquad (11\text{-}1a)$$

11-3 Flow of an Ideal Fluid— Bernoulli's Theorem

An ideal fluid has no viscosity. Unfortunately all real fluids are viscous. Nevertheless it is very useful to consider how flow behaves in the ideal situation, especially when you consider that the ideal situation predicts results that are quite close to results for a real liquid, such as water (with its relatively low viscosity) in moderately large tubes.

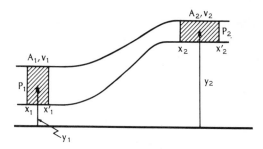

Figure 11.2

Flow of an ideal fluid

Figure 11-2 shows the most general case where the tube changes not only in area but also in elevation. The result during flow will be changes in both the potential and kinetic energies of the fluid. Consequently there will be a total change of mechanical energy of:

$$\Delta E = \Delta E_P + \Delta E_K$$
$$= (mgy_2 - mgy_1) + (\tfrac{1}{2}mv_2^2 - \tfrac{1}{2}mv_1^2) \qquad (11\text{-}3)$$

Some work must have been done to bring about this change. Imagine a particular segment (or slug) of fluid, with its leading and trailing edges initially at x_2 and x_1. As it is pushed forward these boundaries advance to x_2' and x_1'. The fluid behind pushed forward on this segment with a force $F_1 = P_1 A_1$. This force acted through a distance $x_1' - x_1$. Thus the work done on the segment by this force is $W_1 = F_1(x_1' - x_1) = P_1 A_1 \Delta x = P_1 V$. Similarly the fluid in advance of this segment pushed back on it doing work $W_2 = -P_2 V$. Since the fluid is incompressible the two V's are equal. The net work done on the fluid is:

$$\Delta W = W_1 + W_2 = (P_1 - P_2) V$$

also

$$V = m/\rho \quad \therefore \quad \Delta W = (P_1 - P_2)m/\rho \quad (11\text{-}4)$$

Letting $\Delta W = \Delta E$ and re-arranging terms then:

$$\frac{P_1}{\rho g} + \frac{v_1^2}{2g} + y_1 = \frac{P_2}{\rho g} + \frac{v_2^2}{2g} + y_2 \quad (11\text{-}5)$$

Or in general at any point:

$$\frac{P}{\rho g} + \frac{v^2}{2g} + y = \text{a constant} \quad (11\text{-}6)$$

This equation in either form (11-5) or (11-6) is known as "Bernoulli's equation." It could also be written as:

$$P + \tfrac{1}{2}\rho v^2 + \rho g y = \text{a constant} \quad (11\text{-}7)$$

Strictly speaking, the pressure should be an absolute pressure but for most real cases it can also be a gauge pressure. This is because most problems involve only pressure differences.

EXAMPLE

Oil of density 850 kg m^{-3} flows in a tube 0.03 m in diameter at a pressure of 1.6×10^5 Pa. At a smooth constriction the tube diameter reduces to 0.02 m and the pressure to 1×10^5 Pa. Calculate the rate of flow of oil in the tube. Assume the tube is horizontal and that the pressures are absolute pressures.

The Bernoulli equation describing the flow is:

$$\frac{P_1}{\rho g} + \frac{v_1^2}{2g} + y_1 = \frac{P_2}{\rho g} + \frac{v_2^2}{2g} + y_2$$

Since y_1 equals y_2 then the g's cancel and

$$\tfrac{1}{2}(v_2^2 - v_1^2) = \frac{(P_1 - P_2)}{\rho}$$

$$= \frac{0.6 \times 10^5}{850} \text{ m}^2 \text{ s}^{-2} \quad (1)$$

Also $A_1 v_1 = A_2 v_2$

$$\therefore \quad \pi\left(\frac{0.03}{2}\right)^2 v_1 = \pi\left(\frac{0.02}{2}\right)^2 v_2$$

and $v_2 = 2.25 v_1$ (2)

Substitute for v_2 in (1) thus

$$v_1 = 5.9 \times 10^{-3} \text{ m s}^{-1}$$

then the flow

$$Q = A_1 v_1 = \pi\left(\frac{0.03}{2}\right)^2 \times 5.9 \times 10^{-3}$$

$$= 4.2 \times 10^{-6} \text{ m}^3 \text{ s}^{-1}$$

EXAMPLE

By what percentage would the pressure drop in an artery as the blood enters a region which has been narrowed down by an atherosclerotic plaque to a cross-sectional area only $\tfrac{1}{5}$ of normal?

The vessel is assumed to be fairly large so that the error introduced by ignoring the viscosity of the blood is not great. This allows the use of Bernoulli's equation. Within a very short distance any change in elevation will also be negligible. Normal blood pressure in the larger arteries averages about 100 mm Hg. We will use this value as P_1. A typical value for v_1 would be 0.12 m s^{-1}.

$$\frac{P_1}{\rho g} + \frac{v_1^2}{2g} + y_1 = \frac{P_2}{\rho g} + \frac{v_2^2}{2g} + y_2$$

$$\therefore \quad P_1 - P_2 = \frac{\rho}{2}(v_2^2 - v_1^2) \quad (1)$$

also

$$A_1 v_1 = A_2 v_2$$

$$\therefore \quad v_2 = 5 v_1 \quad (2)$$

Substitute for v_2 in (1):

$$P_1 - P_2 = 12\rho v_1^2$$

$$= 12 \times 1000 \text{ kg m}^{-3} \times (0.12 \text{ m s}^{-1})^2$$

$$= 170 \text{ Pa}$$

$P = 100$ mm Hg. To convert to Pa we use $P_1 = \rho g h$

$\therefore P_1 = 13{,}600 \text{ kg m}^{-3} \times 10 \text{ m s}^{-2} \times 0.01 \text{ m}$

$\quad = 1{,}360 \text{ Pa}$

$\therefore \dfrac{\Delta P}{P} = \dfrac{170}{1360} \times 100 = 12$ percent decrease in

pressure in the constricted area.

One of the most important consequences of Bernoulli's law is that whenever a fluid flows faster there is a drop in pressure (as long as there is not a simultaneous considerable increase in elevation of the tube). The decrease in pressure can be calculated quantitatively for an incompressible fluid. Qualitatively the same type of pressure drop occurs when a compressible fluid, e.g. air, is involved. This drop of pressure can be used to explain many phenomena from the lift on a wing to the fact that baseballs hit down the first or third base lines often curve foul. The same phenomena can be explained by Newton's third law of motion as well.

11-4 Flow of a Viscous Fluid— Poiseuille's Law

There are situations where viscous effects in the flow of fluid cannot be ignored. For flow in tubes of small internal diameter, especially for a liquid with a large coefficient of viscosity, Bernoulli's law is no longer applicable. This is particularly true for blood in the smaller vessels of the circulatory system. In the nineteenth century, Poiseuille worked out the details for Newtonian fluids in an effort to measure the viscosity of blood by measuring its flow rates in a cylindrical tube. Water, at a given temperature, has a constant viscosity regardless of the method used to measure it and regardless of the shear forces in the fluid. Any liquid that behaves like this is referred to as "Newtonian." Blood is, to a very close approximation, Newtonian down to very small tube radii but some other liquids, e.g. many paints, are very non-Newtonian.

The derivation of the equation for flow of a viscous fluid in a tube of radius R and length ℓ will be accomplished in two stages. First a formulation is derived for the velocity profile across the tube and second, the resultant flow across the tube is integrated.

Consider a cylinder of very small radius r at the centre of the tube (Figure 11-3).

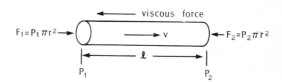

Figure 11.3
Flow of a viscous liquid

A net driving force $F_1 - F_2 = (P_1 - P_2)\pi r^2$ forces fluid to the right but a viscous force acts at the surface of the cylinder to oppose the motion. Unlike the frictional forces encountered between dry solid surfaces, this frictional force is velocity dependent and given by (see Chapter 9):

$$F = \frac{\eta A v}{L} \qquad (11\text{-}8)$$

where η is, of course, the coefficient of viscosity, $A = 2\pi r \ell$ is the surface area of the cylinder, and L, the thickness of the fluid layer should now be measured from the centre out and be represented as dr. Consequently the viscous force has a magnitude of:

$$\eta 2\pi r \ell \frac{dv}{dr}$$

Since v decreases as r increases, a negative sign must be introduced:

$$F = -\eta 2\pi r \ell \frac{dv}{dr} \qquad (11\text{-}9)$$

The velocity, and consequently the viscous force, will increase until the viscous force is equal and opposite to the driving force. At this point, equilibrium is attained and a constant velocity and flow results. Mathematically:

$$-\eta 2\pi r \ell \frac{dv}{dr} = (P_1 - P_2)\pi r^2$$

or

$$-dv = \frac{(P_1 - P_2)}{2\eta\ell} r\, dr \qquad (11\text{-}10)$$

The velocity of flow at any radius r from the centre of the tube can be obtained by integrating this equation over appropriate limits, i.e.:

$$-\int_v^0 dv = \frac{(P_1 - P_2)}{2\eta\ell} \int_r^R r\, dr$$

whence:

$$v = \frac{(P_1 - P_2)}{4\eta\ell}(R^2 - r^2) \qquad (11\text{-}11)$$

This equation predicts a parabolic profile (actually a paraboloid of revolution) decreasing from a maximum at the centre (where $r = 0$) to a minimum of 0 at the outside (where $r = R$).

To calculate the total flow in the tube, recall that $Q = $ flow rate $ = \Delta V/\Delta t$. Picture now looking down the tube and observing an annulus or ring of flow of thickness dr coming towards you (Figure 11-4). Since the annulus is very thin the very small flow in it can be represented as $dQ = dV/dt$. The volume of flow in this annulus is:

$$dV = dA\, d\ell$$

since

$$d\ell = v\, dt$$

and

$$dA = 2\pi r\, dr$$

$$\therefore dV = 2\pi r v\, dr\, dt$$

and

$$dQ = 2\pi r v\, dr \qquad (11\text{-}12)$$

The total flow summed over all such annuli will be

$$Q = \int_0^R 2\pi r v\, dr \qquad (11\text{-}13)$$

Substitution for v from Equation (11-11), followed by integration yields:

$$Q = \frac{\pi R^4 (P_1 - P_2)}{8\eta\ell} \qquad (11\text{-}14)$$

which is called Poiseuille's law. The ratio $(P_1 - P_2)/\ell$ or $\Delta P/\ell$ is called the pressure gradient.

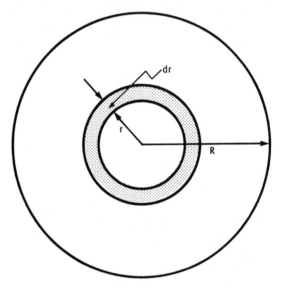

Figure 11.4

Annulus of flow in a tube

It is unusual for one variable, such as Q, to be dependent on the fourth power of another variable (as R). This means that, other factors remaining unchanged, a reduction in the radius to one half the original value will result in a flow of one sixteenth the original value. This can be of physiological significance in the circulation where reduction of arteriole radius due to smooth muscle contraction can change flow rates quite markedly. Such changes occur in response to direct nervous stimulation or to circulating sympathomimetic hormones such as adrenaline (epinephrine).

An arteriole has a diameter of 2.5×10^{-5} m with blood flowing at 2.8×10^{-3} m s^{-1}. What is ΔP from one end to the other if the length of the arteriole is 5×10^{-3} m? What fraction of the total pressure drop in the circulation does the ΔP represent?

$$Q = Av = \pi R^2 v = \frac{\pi R^4 \Delta P}{8 \eta \ell}$$

$$\therefore \Delta P = \frac{8 \eta \ell v}{R^2}$$

assuming $\eta = 4.5 \times 10^{-3}$ N s m^{-2} (typical for blood)

$$\Delta P = \frac{8(4.5 \times 10^{-3} \times 5 \times 10^{-3} \times 2.8 \times 10^{-3})}{(1.25 \times 10^{-5})^2}$$

$$= 3200 \text{ Pa}$$

Normal average blood pressure in the large vessels (therefore approximately in the heart) is about 100 mm mercury. By the return to the heart the pressure is approximately 0. Therefore the entire pressure drop of 100 mm Hg in the circulation is:

$$\Delta P_{(total)} = \rho g h = 13,600 \times 9.8 \times 0.10$$

$$\approx 1.36 \times 10^4 \text{ Pa}$$

Thus the drop in the arteriole is $(3200/1.36 \times 10^4) \times 100 \approx 24$ percent of the total drop in the circulation.

The previous derivation assumed that the velocity profile all along the tube was parabolic. If the tube is relatively short and fed from some larger reservoir, then there are inlet effects and the flow profile may not become fully parabolic until a length equivalent to a few tube radii has been traversed. Figure 11-5 indicates schematically the development of the profile along the tube.

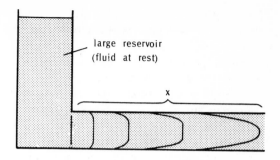

Figure 11.5
Inlet effects in streamline flow

The inlet length for streamline flow has been found empirically to be approximately

$$x = 0.057(R)(R_e) \qquad (11\text{-}15)$$

where R_e (called the Reynold's number) is a flow dependent parameter to be defined in the next section. In the case of arteries where the flow is somewhat pulsatile, the inlet length is approximately

$$x = 0.08(R)(R_e) \qquad (11\text{-}16)$$

For precise results when flow is from a stationary reservoir into a relatively short tube (a hypodermic needle would be an approximate instance) Poiseuille's law must be modified to:

$$Q = \frac{\pi R^4 \left[\Delta P - \dfrac{Q^2 \rho K_2}{\pi R^4} \right]}{8 \eta (\ell + K_1 R)} \qquad (11\text{-}17)$$

where $K_1 \approx 1.64$ and $K_2 \approx 1.0$.

The bulk of water movement in the trunks of trees is through the xylem tissue where long cells, typically 1 mm long, and of radius from 20 μm to 200 μm, depending on the species, are joined more or less end to end to form conducting pathways that approximate cylinders. It might be expected that Poiseuille's law would apply in these vessels. The measurements are difficult to carry out but do seem to be consistent with Poiseuille's law. In trees with vessels of 20 μm radius the observed upward flow velocity is 0.1 cm s^{-1} which would require, if we assume $\eta = 0.001$ N s m^{-2} a pressure gradient ($\Delta P/\ell$) of 2×10^4 Pa m^{-1}. The observed pressure gradient is 3×10^4 Pa m^{-1} which looks like rather poor agreement. However, if the trunk is placed in a horizontal position only the expected 2×10^4 Pa m^{-1} are required. In the vertical position the additional 1×10^4 Pa m^{-1} is required to overcome the gravitational potential.

It was mentioned earlier that blood is slightly non-Newtonian. If we flow water, which is Newtonian, through tubes of different radii and use Poiseuille's law to calculate the coefficient of viscosity, we will get the same value for all tube sizes. If we repeat the same experiment with blood we find that the blood apparently becomes less viscous as the tube radius decreases. This is referred to as non-Newtonian behaviour. The effect is almost undetectable for larger tubes but as the radius decreases below 500 μm (0.5 mm) the phenomenon, known as the Fahreus-Lindqvist effect, can be quite significant (see Figure 11-6).

Newtonian behaviour. Probably a number of phenomena contribute. One of the more important factors is probably the fact that velocity gradients tend to cause the erythrocytes (red blood cells) to tumble in towards the centre of the stream. Consequently, a boundary layer develops next to the walls in which the number of cells is significantly reduced so that the flow here is essentially that of the plasma which has a lower viscosity. The size of this boundary layer relative to the total cross-sectional area becomes relatively more important as the radius of the tube gets smaller.

Often a large tube divides into two or more smaller tubes, such as frequently happens in the circulation. In this case, Poiseuille's law requires that to maintain the same total flow with similar pressure gradients the total cross-sectional area of all n tubes after division is equal to $(n)^{1/2}$ times the area of the original tube. The division into four smaller tubes requires twice the original area and each of the new tubes must be $\frac{1}{2}$ the original area. This is part of the explanation of why the total cross-sectional area of all the capillaries is much greater than the cross-section of the aorta.

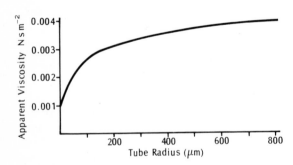

Figure 11.6
Apparent viscosity of blood of hematocrit 40% (i.e. 40% of volume was red blood cells)

There is considerable disagreement as to the complete explanation of this non-

EXAMPLE

A blood vessel of radius R_1 divides into n vessels of equal (smaller) radii R_2. What must be the ratio of the total area of the smaller vessels to the area of the larger vessel if the same pressure gradient ($\Delta P/\ell$) exists in both? How does the velocity of the blood in the smaller vessels compare to that in the larger vessels?

$$Q_1 = nQ_2$$

$$A_1 v_1 = nA_2 v_2$$

$$\therefore \frac{\pi R_1^4}{8\eta} \frac{\Delta P}{\ell} = \frac{n\pi R_2^4}{8\eta} \frac{\Delta P}{\ell}$$

$$\therefore \sqrt{n}\, R_2^2 = R_1^2$$

and

$$\sqrt{n}\,\pi R_2^2 = \pi R_1^2$$

i.e. the original area of the original large vessel equals $n^{1/2}$ times the area of *each* new vessel and the total area of *all* new vessels will be

$$n\pi R_2^2 = \sqrt{n}\,\pi R_1^2$$

Thus if a vessel divides into 2 equal vessels the total area will increase by $2^{1/2} = 1.4$ times. From the equation of continuity,

$$v_2 = v_1/\sqrt{n}$$

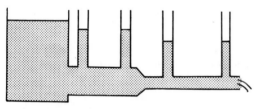

pressures predicted by Bernoulli's equation

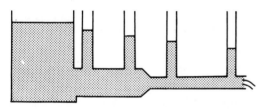

pressures predicted by Poiseuille's law

Figure 11.7
Pressure along a flow tube

It may have been noticed that Poiseuille's law and Bernoulli's equation are contradictory. Poiseuille's law predicts, as a result of viscosity or fluid friction, a pressure drop along a horizontal tube of constant diameter. Bernoulli's equation predicts no pressure drop under the same conditions since it deals with an ideal (and therefore frictionless) fluid. In Figure 11-7 the pressures (as indicated by rise

of fluid in vertical tubes) are compared along the same flow system under the conditions of both equations. The student is frequently uncertain as to which equation should be used. If the fluid is viscous and flowing through a tube which does not change in radius, then Poiseuille's law must be used. If there is a constriction or dilation in the vessel then only Bernoulli's equation can be used even if some error is introduced by neglecting any viscous effects. These errors will be fairly small for liquids with viscosities like blood or water flowing in tubes of moderately large radius, but can be quite large with very viscous liquids and/or tubes of very small diameter.

11-5 Turbulent Flow

If a greater and greater pressure difference is applied to a tube through which a Newtonian fluid is flowing, the flow will increase linearly, in accordance with Poiseuille's law, but only up to a certain point beyond which it takes significantly greater increments of pressure to produce flow increases (see Figure 11-8).

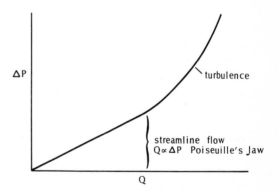

Figure 11.8
Flow in a tube

Beyond the Poiseuille law region, the flow will be definitely noisy and the tube may even vibrate. Ink injected as before would be observed to swirl around rather than advance in streamlines. The flow is said to be turbulent.

In general turbulant flow is inefficient and should be avoided if possible.

The forward components of the velocities of various particles in turbulent flow form a rather flat profile (Figure 11-9). The very narrow zone along the walls in which velocity is changing most rapidly is frequently a boundary layer of streamline flow.

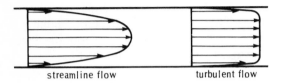

streamline flow   turbulent flow

Figure 11.9
Velocity profiles

Many years ago Reynolds investigated turbulent flow and discovered that a dimensionless parameter, now known as Reynolds number (R_e), could be very useful in characterizing flows.

$$R_e = \frac{\rho v D}{\eta} \qquad (11\text{-}18)$$

where D is the diameter of the tube, and v is the average flow velocity.

Frequently, the flow in simple liquids such as water or blood would change from streamline to turbulent at Reynolds numbers around 2000 to 2500. Unfortunately a wrong impression has been created by much of the literature and people are often left with the impression that flow will be turbulent above a critical velocity which is calculated from Equation (11-18) when Reynolds number is set at about 2000. This is not correct! With great care and very smooth walled tubes, it is quite possible to have streamline flow when Reynolds number is as great as 16,000. Even in models of the aorta with its curvature and side branches streamline flow has been obtained at $R_e \approx 16,000$. The value of 2000 is neverthe-

less of some importance since whenever there is a disturbance at $R_e > 2000$ it will usually propagate and turn into full turbulence whereas a disturbance at $R_e < 2000$ will usually die away and streamline flow will be restored. Such a disturbance might occur at an imperfection in the tube wall.

Some books define Reynold's number using radius instead of diameter. In these cases the 2000 figure we have been using is changed to 1000. If an R_e value is quoted in a book or set of tables always check to find out whether diameter or radius was used in the calculation.

EXAMPLE

What is the peak Reynold's number in the abdominal aorta of the rabbit where $D = 0.3$ cm and the peak velocity $v = 0.60$ m s^{-1}. If a disturbance is initiated will it propagate or die out?

$$R_e = \frac{\rho D v}{\eta} = \frac{1000 \times 3 \times 10^{-3} \times 0.60}{0.004}$$

$$= 450$$

Since R_e is less than 2000 then the disturbance will probably die out.

EXAMPLE

Would you expect turbulence in the human aorta? ($D \approx 0.02$ m, $\rho = 10^3$ kg m^{-3}, peak velocity $= 0.40$ m s^{-1} (for a man at rest), $\eta = 0.004$ N m^{-2} s)

$$R_e = \frac{\rho v D}{\eta} \approx \frac{10^3 \times 0.40 \times 0.02}{0.004} = 2000$$

This is so close to the traditional "dividing point" that you cannot answer the question in a definite manner. Direct observation suggests that little if any turbulence occurs in the aorta even when the flow increases as in heavy exercise. There may be a little turbulence right at the peak of systole (ejection of blood from the heart) especially at the valves but if so, it quickly dies out. In no other vessel is the

Reynolds number this large. Consequently there is essentially no turbulence in the blood vessels of a healthy person with the exception of local disturbances at bifurcations (branching points) which quickly die out within a very short distance.

11-6 Bolus Flow

Many capillaries have a diameter of only 5 or 6 μm whereas the erythrocytes have a diameter of about 8 μm. Consequently the erythrocyte must deform to pass through the capillary. The cells occupy the entire cross-section of the interior of the vessel and pass through as a series of moving plugs with short sections of plasma trapped between them. This is known as bolus flow and is represented schematically in Figure 11-10.

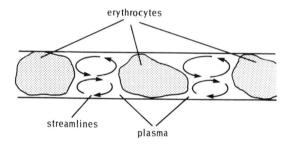

Figure 11.10
Bolus flow

The flow in the trapped plasma sections becomes a specialized form of streamline flow. Rapidly moving streamlines down the centre catch up with an erythrocyte and are deflected to the outside where their velocity, relative to the wall, becomes essentially zero. The following erythrocyte picks up these layers and forces them once more down the centre. The streamlines advance much as the tread on a bulldozer. This has a decidedly beneficial result as the plasma sections are kept well stir-

red up and this facilitates rapid movement across the capillary walls of waste products, etc.

11-7 Pulsatile Flow

In the larger arteries and the thoracic veins the beating action of the heart creates a pulsatile flow. The blood still advances in streamlines but the shape of the velocity profile is not parabolic. The profile will be rather square across most of the area but changes in magnitude throughout the cycle with the flow actually reversing at one point in the cycle. If the velocity profile is examined over several cycles, one gets the impression of a slug of fluid which advances four steps and then backs up one and repeats this over and over. As we proceed away from the heart the pulses are gradually damped out due to several factors including the viscosity of the blood and the extensibility of the blood vessel walls. In the smaller arteries, if average values are used for the pressure gradient, Poiseuille's law is quite satisfactory for predicting the average flow. In the larger vessels however the observed flow will be much less than that predicted by Poiseuille's law, and may be as little as one-fifteenth of that predicted.

Figure 11-11 shows the pulsatile pressure and flow rate in the femoral (thigh) artery of a dog. The heart rate is $2.75\,\text{s}^{-1}$ and the period is represented on the horizontal axis as 360°. Notice that the flow reverses for part of the cycle even this far from the heart. The total flow during the cycle could be obtained by integrating the area under the curve Q. It might seem strange that flow will reach its maximum before the pressure reaches its maximum. However, it is not pressure that causes flow but the pressure gradient. The pressure gradient is the derivative of the pressure curve and the maximum pressure gradient occurs just in advance of the maximum flow.

The curves shown in Figure 11-11 are both

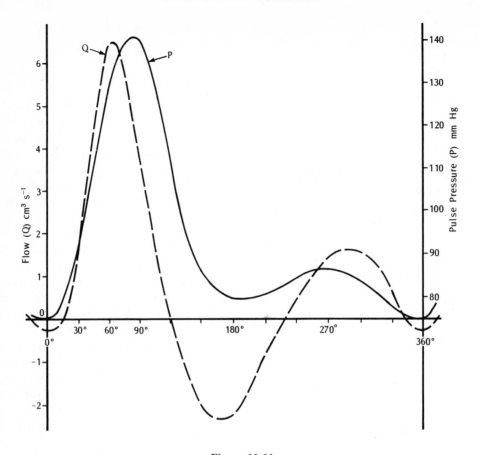

Figure 11.11

Pulsatile pressure and flow in the femoral artery of the dog (from D. A. McDonald "Blood Flow in Arteries", Edward Arnold, Publishers 1960)

experimentally determined. It is moderately difficult to measure the pressure curve and very difficult to measure the flow curve. Is it then possible to predict the flow from the pressure? It is but the mathematical techniques are rather advanced. We cannot, in this book, solve the problem quantitatively but we can discuss qualitatively how it is accomplished.

Any periodic function such as the pressure gradient curve can be broken down into a series of simple sine curves of different amplitudes and phases (see Appendix 3) which are the first, second, third, etc. harmonics of the fundamental frequency. The process by which this is accomplished is known as Fourier analysis. In this particular case, a series of four terms is adequate. For each sinusoidal pressure gradient component, a sinusoidal flow component can be calculated. Simple addition of the four flow components gives a very precise representation of the total experimentally measured flow as presented in the original graph.

11-8   Development of an Aneurysm

We have tended to ignore the blood vessel itself treating it as a rather passive tube through which flow occurs. The artery walls possess quantities of the structural proteins elastin and collagen as well as contractile smooth muscle cells. The relative amounts of these components vary with distance from the heart. In the largest vessels, elastin and collagen are the most important contributors to the elastic properties of the wall. A graph of tension in the wall versus radius of such a vessel is shown in Figure 11-12. The lower portion is principally due to the elastin component while the more steeply rising portion is due principally to the collagen component. In the smaller arteries, there is considerable smooth muscle, the active contraction of which can introduce a tension component in addition to the passive elastic components shown in Figure 11-12. A number of interesting computations can be performed on the equilibrium of these various vessels. We shall restrict ourselves to one particular problem that can develop in the larger arteries where

we do not have to worry about the complicating contribution from smooth muscle.

Across any curved flexible liquid surface there is a pressure difference (Chapter 9):

$$\Delta P = \gamma \left(\frac{1}{R_1} + \frac{1}{R_2} \right) \qquad (11\text{-}19)$$

where the greater pressure is on the concave side. In the case of a blood vessel the surface tension must be replaced by the tension (T) in the wall. The radii describe the curvature at any point. They are minimum and maximum radii of orthogonal arcs at any point. For a cylinder of radius R, the other radius will be that of a straight line along the surface. Thus one of the radii goes to infinity and its reciprocal to zero. Consequently Equation (11-19) can be re-written for the blood vessel wall as:

$$T = \Delta P R \qquad (11\text{-}20)$$

This is the equation of a straight line of slope ΔP, passing through the origin. In Figure 11-12 we show such lines for two pressure differences ΔP_1 and ΔP_2. For each ΔP the abscissa of the point where the straight line intersects the T vs. R curve fixes the radius of the vessel.

It is conceivable that ΔP might increase until equal to the slope of the T vs. R curve. There would be no point of intersection and R could increase in an unrestricted manner, perhaps even causing the vessel to rupture. This type of "blow-out" can and does occur and is known as an aneurysm (which all too often has fatal results).

When will this occur? When the ΔP reaches some maximum value ΔP_{max} such that:

$$\Delta P_{max} = \frac{dT}{dR}\bigg|_{max} \qquad (11\text{-}21)$$

The equation defining Young's modulus (Chapter 7) can be re-written as;

$$\frac{F}{A} = Y \frac{\Delta \ell}{\ell} \quad \text{becomes} \quad \frac{dT}{t} = Y \frac{dR}{R_o}$$

$$(11\text{-}22)$$

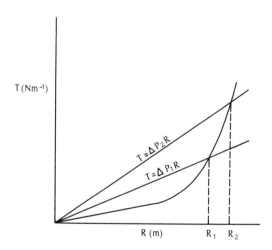

Figure 11.12

Tension versus radius in a healthy elastic artery

where t is the thickness of the wall and R_o is the "normal" radius of the vessel.

The maximum value of dT/dR then will be:

$$\left.\frac{dT}{dR}\right|_{max} = \frac{Y_{max}t}{R_o} = \Delta P_{max} \qquad (11\text{-}23)$$

Experimentally it has been found that Y_{max} in healthy human iliac arteries is 7×10^5 N m^{-2} at age 30 and 1.8×10^6 N m^{-2} at age 80. R_o is 0.35 cm and $t = 0.07$ cm. Substitution of these values (at age 30) into Equation (11-23) shows that the ΔP required for an aneurysm is 1.4×10^5 N m^{-2}, considerably more than one atmosphere. This is equivalent to about 1000 mm of mercury, whereas peak systolic pressure is normally only about 120 mm of mercury. The heart cannot develop such a pressure. An aneurysm therefore cannot develop in a healthy vessel but only one where something, such as a connective tissue disease, has led to a significant reduction of the maximum value of Young's modulus in the blood vessel wall.

Problems

1. Which of the following statements is not implied by the Bernoulli equation?
   (a) In a horizontal tube, if the velocity increases the pressure decreases.
   (b) In a tube of variable cross-section, the velocity times the area is a constant.
   (c) The pressure increases with depth in the ocean.
   (d) The total energy per unit volume of a fluid is constant.
   (e) The kinetic energy may increase at the expense of the gravitational energy.
2. Suppose that the flow of blood were to stop in the leg (of length 0.76 m). What is the pressure of the blood in the foot rela-tive to that at the hip, if the density of blood is that of water and if the patient remains standing?

3. An ideal fluid flows in the tube which constricts and drops. What must h be in order that the pressure in the fluid at the bottom (P_2) equal the pressure at the top (P_1)?

4. If the absolute pressure in the larger section of the cylindrical tube shown is 2.0×10^5 Pa, then what is the absolute pressure in the smaller section? (Assume that the liquid is water and that its viscosity is negligible.)

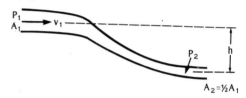

5. The area of a pipe decreases linearly from A to $\frac{1}{2}A$ over a distance of 0.15 m. If the speed of the fluid (non-viscous) is 0.10 m s^{-1} and the gauge pressure is 50 Pa at the large cross-section, how high will the fluid be in the tube located 0.10 m downstream? The fluid density is 10^3 kg m^{-3}.

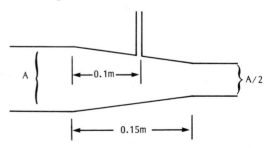

6. Diagrams I and II represent schematically a human leg in the vertical and horizontal positions respectively. If the flow rate is to be the same in both cases:
   (a) The arterial pressure at the entry to the leg must be the same in both cases.
   (b) The arterial pressure at the entry to the leg must be greater in I.
   (c) The arterial pressure at the entry to the leg must be greater in II.

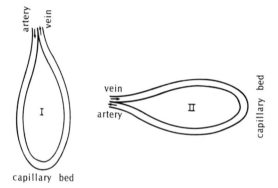

7. The blood in an artery of radius 5×10^{-3} m, flows with a speed of 0.15 m s^{-1}. This artery sub-divides into a large number of capillaries of radius 5×10^{-6} m. The flow velocity in the capillaries is 5×10^{-4} m s^{-1}. Into how many capillaries does this artery divide? (Treat blood as an ideal fluid.)

8. Four small veins all of the same radius R_s join together to form a larger vein of radius $R_l = 5R_s$. The average velocity of the blood in each of the small veins is v_s. What is the average velocity in the larger vein?

9. Poiseuille's law states that $Q = \dfrac{\pi R^4 \Delta P}{8 \eta \ell}$. If ΔP, η and ℓ are held constant, draw graphs Q vs. R and Q vs. R^4. If R, ΔP and η are held constant, draw graphs of Q vs. ℓ and Q vs. $1/\ell$.

10. Many clinical studies reveal that deposits in arteries effectively narrow the arterial passages. Compare a healthy artery and a diseased one whose cross-sectional area is only 0.6 of the healthy one. What change in the pressure drop per unit length is necessary if the same volume of blood is to be carried by the diseased artery per unit time?

11. What is the peak Reynolds number in the abdominal aorta of the rabbit where the diameter $= 0.3$ cm and the peak velocity $= 60$ cm s^{-1}? If a disturbance is initiated in this vessel will it likely propagate or die out?

12. Water at room temperature, flows through a tube of internal diameter 2 cm at an average speed of 20 cm s^{-1}. Which one of the following statements about the flow is correct?
    (a) The flow will possibly be turbulent.
    (b) The flow will undoubtedly be stream-line.
    (c) The flow will undoubtedly be turbulent.
    (d) Any disturbances in the flow will quickly be dampened out.
    (e) The flow will be turbulent if the viscosity of the liquid appears to decrease as the flow is increased from 0 to the final value of 20 cm s^{-1}.

13. (a) Under the conditions of normal activity an adult inhales about 1 litre of air during each inhalation. With a watch, determine the time for one of your own inhalations (average several), and hence calculate the volume flux in m^3 s^{-1} and average linear air speed in m s^{-1}) through your trachea during inhalation. The radius of the trachea in adult humans is approximately 10^{-2} m.
    (b) From the data above, calculate the

Reynolds number for the air flow in the trachea during inhalation. Do you expect that the flow is streamline or turbulent?

14. What is the tension in an arteriole wall if the mean blood pressure is 60 mm Hg and the radius is 0.01 cm?

15. What is the tension in a capillary wall if the blood pressure is 30 mm Hg and the radius 4 micrometres?

16. From the graph of pulsatile flow in the text (Figure 11-11) calculate the approximate flow (a) for each beat of the heart and (b) in one minute.

Chapter 12

Heat and Heat Flow
in Biological Systems

12-1   Introduction

Fish are poikilotherms (cold blooded creatures) and we normally expect their body
temperatures to be essentially that of their
environment. This is true for most fish which
consequently are rather sluggish at low temperatures and more active at higher temperatures. There are, however, some ocean fish,
such as the bluefin tuna which are powerful
swimmers and which, while still poikilotherms,
have developed a special mechanism, the rete
mirabile, for keeping their main swimming
muscles at a temperature several degrees
higher than ambient temperature. In this
chapter we will consider the several aspects of
heat and heat flow required to fully understand this and other fascinating phenomena of
biological heat regulation.

12-2   Heat Capacity

When any quantity of material is at a temperature different from its environment, it will
gain or lose heat energy in an attempt to
achieve an equilibrium or steady state. The
processes by which the heat is exchanged will
be considered in a later section. The quantity
of heat, H joules, required to change the
temperature of m kilograms of a substance by
ΔT kelvin or degrees Celsius is:

$$H = mC\,\Delta T \qquad (12\text{-}1)$$

The specific heat, $C\,\mathrm{J\,kg^{-1}\,K^{-1}}$, which is
characteristic of the substance and its state, is
essentially constant over the temperature
ranges encountered in biological systems.
Water, for instance, has a specific heat of
$4186\,\mathrm{J\,kg^{-1}\,K^{-1}}$ but the value for ice is only
one half of this. A mixture of water vapour
and air on the other hand has a very variable
specific heat which varies with temperature
and pressure. All else being equal it will require more energy to heat a quantity of gas at
constant pressure P than at constant volume
V because in the constant pressure case,
energy must be provided not only to increase
the temperature but also to perform the work
of expansion ($W = P\,\Delta V$) that will simultaneously occur.

In the past it was usual to specify heat
energy in calories rather than joules where
one calorie is the heat required to raise one
gram of water one degree Celsius. One joule
is 0.2388 cal. The energy content of foods has
been historically measured in Calories (note
the capital C) where one Calorie is 1000
calories or 4186 joules. The energy content of
many foods, to the dismay of those with too
large waists, is quite high, i.e. one kilogram of
cheddar cheese can provide 400 Calories or
1.7×10^6 joules. Eventually all our over-
weight Calorie counters will have to become
joule counters but this is one area where the
traditionalists are putting up a firm rear-guard

action and the conversion may be slow.

Not surprisingly cells and tissues of living systems have specific heats rather close to water. The complete body of a mouse, for instance has a specific heat of about $3450 \, J \, kg^{-1} \, K^{-1}$. The corresponding value for aluminum is 886 and for crown glass is 670.

Changes of state are sometimes of importance in biological systems as when water freezes on a leaf or when sweat evaporates from skin. When water at 0°C freezes, it gives up its latent heat of fusion, $L_f \approx 3.35 \times 10^5 \, J \, kg^{-1}$ before any additional drop in temperature occurs. This is 80 times the heat given up in cooling the same mass of water one degree Celsius. When water at 100°C is converted to steam, the latent heat of vaporization $L_v = 2.26 \times 10^6 \, J \, kg^{-1}$ must be provided. To vaporize perspiration from your skin, which is at a temperature lower than 100°C, would require even more energy, approximately $2.4 \times 10^6 \, J \, kg^{-1}$. If this heat is provided by the body, it can have a significant cooling effect on the body. Even when human skin is not wet from perspiration there is a continuous loss of water by evaporation. This is called insensible perspiration. In addition the air we breathe out normally has a much higher moisture content than that which we breathe in. These two evaporative losses account, very approximately, for sixteen and eight percent respectively of the total heat loss in man. The heat loss in warming up inspired air is negligible except in incredibly cold weather. Many animals, such as the dog, do not perspire and must pant when overheated.

In any isolated system there must be conservation of energy. While heat is only one of many forms of energy it alone is frequently conserved. Thus if a cold rock or fish were placed in an insulated aquarium with some warm water an exchange of heat would occur and all components would quickly reach an equilibrium temperature. On the other hand, if a scuba diver enters cold water it will be a long time before his deep body temperature changes significantly since he will be converting large amounts of stored chemical energy to thermal energy. If the water is quite cold, this conversion will be unable to keep up with losses to the environment and the diver will eventually start to experience a drop in deep body temperature.

EXAMPLE

A 1 kg fish ($C \approx 3500 \, J \, kg^{-1} \, K^{-1}$) at a temperature of 10°C is placed in 10 kg of water at 20°C. Assuming the system to be thermally insulated and that the contribution from metabolic activity in the fish is negligible, what is the final temperature, T_f, of the fish and the water?

Heat gained by the fish + heat lost by the water = 0.

$$\therefore m_{fish} \times C_{fish} (\Delta T_{fish}) + m_{H_2O} \times C_{H_2O} (\Delta T_{H_2O}) = 0$$

$$\therefore 1 \times 3500(T_f - 10) + 10 \times 4186(T_f - 20) = 0$$

from which $T_f \approx 19.2°C$.

EXAMPLE

If a 1 kg block of ice at 0°C were added instead of the fish what would be the final temperature?

Here the ice will melt, obtaining its latent heat of fusion from the water, and then warm up.

$\therefore$ Heat gained by ice in changing from ice at 0°C to water at 0°C + heat gained by the ice water + heat lost by the original water = 0.

$$\therefore m_{ice} \times L_f + m_{ice} \times C_{H_2O}(\Delta T_{ice \, water})$$
$$+ m_{original \, H_2O} \times C_{H_2O} (\Delta T_{original \, water}) = 0$$

$$\therefore 1 \times 3.35 \times 10^5 + 1 \times 4186(T_f - 0)$$
$$+ 10 \times 4186(T_f - 20) = 0$$

which is solved to yield $T_f = 10.9°C$.

12-3 Heat Exchange

The three main processes by which a body may exchange heat with its surroundings are conduction, convection, and radiation.

In conduction the rate of thermal energy flow, P watts (joule per second) is given by

$$P = kA \, \Delta T/\ell \qquad (12\text{-}2)$$

where $A(\text{m}^2)$ is the cross-sectional area and $\ell(\text{m})$ the length of the object, ΔT is the temperature difference across ℓ. The coefficient of thermal conductivity, k, has units of $\text{W m}^{-1}\text{K}^{-1}$. Some typical values are listed in Table 12-1. Note the similarity of this equation to that of diffusion (Equation 10-15) discussed in an earlier chapter.

Table 12-1
Thermal Conductivities $(\text{W m}^{-1}\text{K}^{-1})$
for a Number of Substances

Substance	k
Aluminum	201
Ice	2.1
Cork	0.05
Fat	0.2
Muscle	0.4
Skin	0.3

The last three figures in Table 12-1 are approximate values for tissues with no blood flow. With blood flow there is a physical transfer of a heated liquid into the tissues (this is analogous to forced convection) and consequently heat flows may deviate markedly from those predicted by Equation (12-2). The blood flow in the skin can vary from as much as 170 kg of blood per second per kilogram of skin down to 1/100 of this value.

Convection is a process by which the physical movement of a fluid (liquid or gas) can transfer heat energy from one location to another. On a global scale this is most evident in many characteristic air and sea currents, such as the gulf stream, that prevail in different areas. Local heating of a mass of fluid, as at the equator, will cause a decrease in density of the fluid which will rise and be displaced away from the equator by colder, denser fluids moving in from below. The rotation of the earth causes a deviation to the right in the northern hemisphere and an opposite deviation in the southern. Thus the combination of convection and the earth's rotation tends to keep fluid masses rotating in a clockwise direction in the northern hemisphere and counter-clockwise in the southern.

As mentioned previously, the circulation of the blood is forced convection and is very effective in transferring deep body heat to our extremities on cold days—though sometimes not as efficiently as we might wish. Wind passing over a body creates another forced convection which can certainly cause impressive heat losses. For a well-clad human on a cold day, the heat loss will be proportional to the temperature difference between himself and his environment and also roughly proportional to the wind speed. For a naked man on a cold day (perish the thought!) the heat loss will be severe in a gentle breeze but would not get much worse in a strong wind because the rate would be limited by body processes.

If a person must work outdoors in cold weather his hands may become "cold-adapted." This means that the average blood flow will be decreased thus lowering the average temperature of the hands and decreasing heat loss. The exposed head, however, cannot adapt like this as the brain must obviously be kept warm to function properly. Consequently losses from an uncovered head on even a moderately cold day can be equal to the loss from all the rest of a well covered body. In spite of this, many people who would always close the door to conserve heat go around all winter with their heads uncovered.

A simple equation was sufficient to calculate the heat flow in conduction. Unfortunately convection is complicated and no single equation prevails. The nature of the surface (flat or curved), the orientation of the surface (horizontal or vertical) the nature of the fluid (gas or liquid) in contact with the surface and the physical properties of the fluid (viscosity, etc.) all have to be considered.

In general, all that can be said is that the heat lost per unit time by convection will be:

$$P \propto A \, \Delta T \qquad (12\text{-}3)$$

where A is the surface area and ΔT the difference in temperature between the surface and the adjacent fluid. The constant of proportionality that would be introduced would be a complex function of all the parameters mentioned in the previous paragraph and, in many instances, would have to be determined empirically. This type of heat loss can be responsible for most of the heat lost from a window on a cold day when the inside and outside of the glass are almost at the same temperature which is approximately half way between the exterior and interior temperatures. The heat, of course, traverses the glass according to Equation (12-2) but the ΔT across the glass is quite small.

Heat energy can also be transmitted or received by the process of radiation in which electromagnetic energy may travel through a vacuum or a transparent medium. In earlier chapters it was pointed out that certain compounds (e.g. the chlorophylls) could absorb such radiation in preferred wavelength bands, or emit it in preferred portions of the spectrum such as in fluorescence. In those instances the concern was with the visible and near visible portions of the spectrum. As far as heat energy is concerned the vast bulk of the radiation is in the infrared portion of the spectrum and most animals, including man, are "black bodies" as far as such radiation is concerned. This means that they are effective radiators or absorbers of all wavelengths within the spectral region of interest. There is some variation away from the behaviour of a perfect black body but the deviations are not of great significance.

Planck has determined that, for a black body, the spectral emittance, W_ν watts m^{-2} per unit range of frequency, is given by:

$$W_\nu = \frac{2\pi h}{c^2} \frac{\nu^3}{e^{h\nu/kT} - 1} \qquad (12\text{-}4)$$

where h is Planck's constant, c the speed of light, T the absolute temperature, k the Boltzmann constant, e the base of natural logarithms (2.718...) and ν the frequency of radiation. This relationship is shown in Figure 12-1 where it can be seen that the total energy radiated, as represented by the area under the curve, increases rapidly with temperature and that the frequency ν_m, at which maximum spectral emittance occurs also increases with temperature.

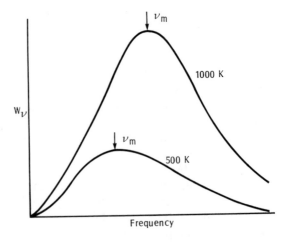

Figure 12.1
Spectral emittance as a function of frequency for two temperatures

The frequency of maximum spectral emittance is given by Wien's displacement law:

$$\nu_m = 5.89 \times 10^{10}\, T \qquad (12\text{-}5)$$

This shift of frequency with increasing temperature explains why an object on heating will become red hot, then white hot, and finally blue hot.

The total intensity radiated, I in $W\, m^{-2}$ for a black body, is given by Stefan's law:

$$I = \sigma T^4 \qquad (12\text{-}6)$$

where

$$\sigma = 5.672 \times 10^{-8}\, W\, m^{-2}\, K^{-4}$$

For a non-black body, the right hand side of the equation must be multiplied by a positive factor which approaches a maximum value of one as the object approaches the behaviour of a black body.

EXAMPLE

Calculate ν_m and I for the two curves in Figure 12-1 at 500 K

$$\nu_m = 5.89 \times 10^{10} \times 500$$

$$= 2.945 \times 10^{13}\, s^{-1}$$

this corresponds to a wavelength

$$\lambda = c/\nu$$

$$\therefore \lambda = 3 \times 10^8\, m\, s^{-1}/2.945 \times 10^{13}\, s^{-1}$$

$$\approx 10^{-5}\, m$$

$$= 10^4\, nm \text{ which is well into the infrared}$$

and

$$I = 5.672 \times 10^{-8}\, W\, m^{-2}\, K^{-4}\, (500\, K)^4$$

$$= 3545\, W\, m^{-2}$$

at 1000 K

$$\nu_m = 5.89 \times 10^{10} \times 1000$$

$$= 5.89 \times 10^{13}\, s^{-1}$$

and

$$I = 5.672 \times 10^{-8}\, W\, m^{-2}\, K^{-4}\, (10^3\, K)^4$$

$$= 5.672 \times 10^4\, W\, m^{-2}$$

The energy radiated per unit area in the previous example seems very large indeed. In practice the net radiation from a black body will rarely approach values of these magnitudes since the same black body will be receiving radiant energy from surrounding bodies according to the same fourth power law. Thus if a body is at one temperature, T_1, which is greater than the surroundings at T_2 it will radiate a net amount of heat:

$$I_{net} = \sigma(T_1^4 - T_2^4) \qquad (12\text{-}7)$$

In real situations, heat exchange usually involves some combination or all of conduction, convection (free and forced) and radiation. Under any given set of conditions the heat exchanged will be proportional to the area and the temperature difference. If the hotter body is that of an animal, then its metabolic activity may allow it to maintain a constant temperature, with a subsequent constant rate of heat loss to its environment if the environment itself does not change. On the other hand, a hot object which contains no internal source of heat, will usually cool off according to Newton's law of cooling. This is an empirical relation which states that the rate of change of temperature will be proportional to the temperature, i.e.:

$$\frac{dT}{dt} = -k(T - T_s)$$

where T_s is the temperature of the surroundings. This can be rearranged and integrated over appropriate limits:

$$\int_{T=T_o}^{T} \frac{dT}{T - T_s} = -k \int_{t=0}^{t} dt$$

The solution of this is:

$$T = T_s + (T_o - T_s)e^{-kt} \qquad (12\text{-}8)$$

The dependence of heat loss on surface area becomes relatively more important for smaller animals since they have a larger surface to volume ratio than large animals. A 0.16 kg pigeon, for instance, loses in a day about $5.2 \times 10^5 \, \text{J kg}^{-1}$ while the corresponding figure for a 680 kg steer is only about one tenth of this.

There are obviously situations where it is desirable for an animal to be able to lose heat, such as during exercise on a hot day, and other cases where the animal should conserve heat, as on a cold day, or in the case of the fish mentioned in the introduction to this chapter. Some animals have been more successful in developing appropriate mechanisms than others.

The rabbit uses its ears extensively for heat regulation. They have a large surface area for good exchange with the environment and specialized connections, called arteriovenous anastomoses, between small arteries and veins. These can open up when the animal is overheated allowing for greater blood flow than could be obtained through capillaries alone. In extreme cold the arteries themselves will constrict and the arteriovenous anastomoses will close down markedly restricting blood flow and consequent heat loss. If the rabbit is fat, the surface to volume ratio will decrease and it will consequently be less able to achieve an adequate rate of heat loss in hot weather or if overheated due to a chase. The northern rabbit will not have the overheating problems of his cousin to the south. There is a gradual decrease of rabbit ear size as one goes from south to north.

In the rat, the tail has developed as a heat exchanger which functions in essentially the same manner as the rabbit ear. There is at least one case on record where rats living in the loft above a hen-house in hot summer weather would sit around with their tails hanging through knot holes in the floor into the cooler environment below, thus increasing their heat loss.

A surprising consequence of the dependence of heat loss on surface area is the fact that insulating very small pipes can actually increase heat losses because the increased surface area can more than compensate for the added insulation. This may be one of the reasons why you have never seen an ant wearing stockings in cold weather.

In the bluefin tuna, heat is retained within the body. Consequently muscle temperatures are higher and this may permit greater swimming speeds. If the blood passed directly from the heart to the gills it would quickly lose to the environment any deep body heat it had acquired. Instead, the vessels leading to the gills break up into a network of fine vessels which intertwine with a similar network of fine vessels returning from the gills. This is called the rete mirabile (literally: the marvellous net). This system acts as a heat exchanger. The cold blood coming from the gills is warmed by the blood going to the gills and returns much of the heat to the deep swimming muscles, rather than allowing it to proceed to the gills to be lost to the environment. Heat exchangers can be built as parallel flow systems where the hottest and coldest fluids are brought together at one end and both emerge as warm fluids at the other end or as counter-flow systems in which the hot and cold fluids enter at opposite ends and pass each other. Engineers have found that this second system is the more efficient. The bluefin tuna apparently "discovered" the same thing millions of years earlier.

Problems

1. While sitting and resting, metabolic processes generate heat in man at a rate of about $60 \, \text{W m}^{-2}$. If all of the heat from a square metre of skin could be transferred to one kilogram of water initially at 20°C, how

long would it take to bring the water to a boil? (The surface area of a man is about 1.5 m^2).

2. Would it require more, less or the same heat to raise the temperature of an aquarium by 10°C when the aquarium is full of water only rather than when filled with water and fish?

3. It has been found experimentally that the heat lost per second by convection from a window when the window is 20°C warmer than the outside air and when there is no wind, is:

$$Q \approx 3.75A \, \Delta T$$

where Q is in watts when A is in m^2 and ΔT in degrees kelvin. The coefficient of thermal conductivity of the glass is 0.84 W m^{-1} K^{-1}. What is the temperature difference across a window 3×10^{-3} m thick when the outside surface is at a temperature of 0°C and the outside air at a temperature of −20°C?

4. What is the temperature of an object whose ν_m is in the middle of the visible spectrum?

5. A black body has a temperature of 5000 K. Compute the ratio of its spectral emittance at 5×10^{14} s^{-1} (in the visible) to its spectral emittance at 3×10^{14} s^{-1} (in the infrared).

6. A cup of hot beverage is initially at a temperature of 100°C. It is located in a room at 25°C. Ten minutes later its temperature has decreased to 80°C. How much longer will it take to cool off to 50°C?

7. A 70 kg man when working hard might generate 230 watts of heat over and above the heat generated while resting. Assuming all of this excess heat generated by the hard work in one hour was removed by the evaporation of water, how much water would be evaporated?

Chapter 13

Biophysics of the Neural Spike

13-1   Introduction

We often correlate the relative positions of organisms on the evolutionary ladder with their relative abilities to receive and process information from their environment. These abilities are directly related to the relative complexity of their sensory and nervous systems. Basic to these nervous systems is their function, the controlled flow of information from one part of the organism to another. The information is in the form of electrical signals, called neural spikes, which arise from precisely determined movements of ions through membranes of specially designed cells. While the field of electronics is primarily concerned with the flow of electrons, not ions, there are certain aspects of the subject which are of great value in helping us understand the mechanisms involved in the production and propagation of the neural spike.

13-2   Electrical Potentials at Membranes

Neurons, the cells which carry the neural spike, are highly specialized for the purpose. The obvious features of one such cell, are shown in Figure 13-1.

Most incoming signals (often chemical in nature) are received at the dendrites which are fairly long fibre-like projections from the cell body. The signals are converted into the neural spike in a part of the cell body known

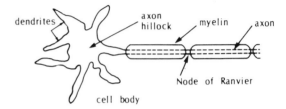

Figure 13.1
The neural cell

as the axon hillock. The spike then propagates itself down the axon, which is another projection of the cell body, usually much longer than the dendrites. It's length is often measured in metres, a distance which is comparable to the lengths of arms or legs of even the largest species.

As is the case with all biological cells, the neuron is surrounded by a membrane about 5 nm thick, even throughout the length of the axon. Most of the axons in the more advanced organisms, such as mammals, are further enwrapped by several double layers of membrane from another cell called a Schwann cell. These additional layers of membrane, termed myelin, do not extend continuously down the axon. Instead they are segmented by small gaps, the nodes of Ranvier. In axons with no myelin, such as those found in squid or lobster, the neural spike propagates "like a burning fuse" by mechanisms which will be described in a following section. In myelinated axons, the neural spike appears only at the nodes of Ranvier. In either case, it is the axon membrane which plays the major role.

One function of the membrane is simply to retard the passage of selected ions and molecules. Ions, however, are electrically charged particles and the flow of such particles is really an electrical current. Any current, I, whether the flow of ions through a membrane or electrons in a wire is measured in amperes. An ampere of current, flowing for one second, amounts to one coulomb of electrical charge passing a given point. Each positive or negative charge such as that carried by sodium ions (Na^+) or chloride ions (Cl^-) is 1.6×10^{-19} coulombs.

The concept of current density, I_d, is very useful, especially in membranes where it is directly related to the rate of flow of ions through a unit area;

$$I_d = \frac{\text{Current passing through a membrane}}{\text{Membrane area involved}}$$

$$= \frac{I}{A} \qquad (13\text{-}1)$$

The ability of ions to pass through a membrane depends on the size, length, shape, and chemical properties of any pores which are present. All these characteristics are lumped together into a single parameter, σ, the conductivity of the pore. The membrane conductivity (g_m) is given by

$$g_m = n\sigma \qquad (13\text{-}2)$$

where n is the number of pores per unit area. Conductivity has units of siemen which are equal to inverse ohms (Ω^{-1}). Typical units for g_m are $\Omega^{-1} cm^{-2}$ or $\Omega^{-1} m^{-2}$. In order to find the actual conductance (g) of a particular area (A) of membrane one must use the relationship

$$g = g_m \times A \qquad (13\text{-}3)$$

EXAMPLE

The squid axon has a radius of 0.25×10^{-3} m, and a membrane conductivity of $1.4 \times 10^{-3} \, \Omega^{-1} cm^{-2}$. Find the conductance of one centimetre of squid axon.

The area of the axon is circumference times length

$$A = 2\pi r \ell$$

$$= 2\pi (0.25 \times 10^{-1} \, cm)(1 \, cm)$$

$$= 0.16 \, cm^2$$

From Equation 13-3,

$$g = g_m \times A$$

$$= 1.4 \times 10^{-3} \times 0.16$$

$$= 2.2 \times 10^{-4} \, \Omega^{-1}$$

The values of the membrane conductivity differ for each ion species because each ion has a characteristic size and charge. Ions also differ in the amount of water they bind. Table 13-1 lists some of the common ions and their corresponding g_m values for squid axon membrane. In any real system such as the squid axon, the internal and external ion concentrations are fairly constant and hence total combined ion membrane conductivity can also be quoted.

Table 13-1
Membrane Conductivity Values for Squid Axon Membrane (Resting)

Ion	g_m
Combined ion	$140 \times 10^{-5} \, \Omega^{-1} cm^{-2}$
K^+	$37 \times 10^{-5} \, \Omega^{-1} cm^{-2}$
Na^+	$1.1 \times 10^{-5} \, \Omega^{-1} cm^{-2}$
Cl^-	$30 \times 10^{-5} \, \Omega^{-1} cm^{-2}$

The motion of ions both inside and outside membranes is partly due to electrical forces arising from electrical charges nearby. The force (F) on a particular ion, having charge q_1, is given by

$$F = Eq_1 \qquad (13\text{-}4)$$

where E is the electric field at this ion due to other charges in the vicinity. If there is only one charge (q_2) producing the field (E_2) at the ion then

$$E_2 = \frac{kq_2}{K(r_{1-2})^2} \qquad (13\text{-}5)$$

which, when combined with Equation (13-4), leads to Coulomb's law:

$$F = \frac{kq_1q_2}{K(r_{1-2})^2} \qquad (13\text{-}6)$$

In these equations r_{1-2} is the distance between the electrical charge (q_2) and the ion, k is a constant having a value $9.0 \times 10^9\,\mathrm{N\,m^2\,C^{-2}}$ and K is the dielectric constant, a property of the material between the charges q_1 and q_2. Note that if K is very large, then the field experienced by the ion is relatively quite small. Table 13-2 contains some dielectric constants for some common biological and non-biological substances. Note the very high value for water, a property which in aqueous solution makes the electric fields due to electrical charges fall off very rapidly with distance from the charge.

Table 13-2
Dielectric Constants of
Selected Materials near 20°C.

Air	1.007
Transformer Oil	2.10
*Palmitic Acid	2.30
*Linoleic Acid	2.60
*Stearic Acid	2.30
Beeswax	2.80
Wood	5.0
Membrane (Approximate)	6.0
Casein	6.5
Ethanol	24.3
Water	80.4

* Membrane Components

Equation 13-6 described the force on an ion due to a single charge nearby. This charge could be a charge on a biopolymer or on another ion. The microscopic environment of biological membranes contains large numbers of ions and charged polymers. Thus the net force on any one ion is due to the combination of the electric fields of all these charges. However, each field is a vector quantity (see Chapter 7) and has a direction (due to the sign and direction of the charge) and a magnitude (due to the proximity and size of the charge). Thus the field experienced by any one ion is the vector sum of all the individual fields. To illustrate this suppose we had a distribution of ions as shown in Figure 13-2(a) and desired to find the field at the position X.

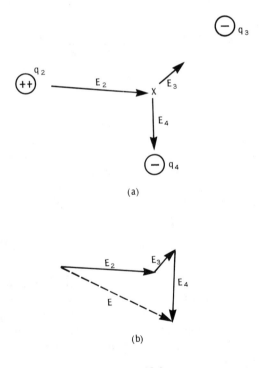

(a)

(b)

Figure 13.2
The electric field due to several charges

The magnitudes and directions of the fields at X due to each charge are shown and then added vectorially in Figure 13-2(b) to produce the resultant field of magnitude E.

EXAMPLE

Find the resultant force on the sodium ion (Na^+) in the environment of calcium ions (Ca^{++}) and chloride ions (Cl^-) shown below. All ions lie in the same plane and the solvent is water.

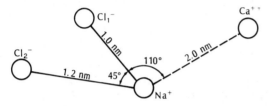

The magnitudes of the individual fields are (using Equation (13-5)):

$$E_{Ca^{++}} = \frac{9.0 \times 10^9 (2 \times 1.6 \times 10^{-19})}{80.4(20 \times 10^{-9})^2}$$

$$= 9.0 \times 10^6 \, N \, C^{-1}$$

$$E_{Cl_1^-} = \frac{9.0 \times 10^9 (-1.6 \times 10^{-19})}{80.4(10 \times 10^{-9})^2}$$

$$= 17.9 \times 10^6 \, N \, C^{-1}$$

$$E_{Cl_2^-} = \frac{9.0 \times 10^9 (-1.6 \times 10^{-19})}{80.4(12 \times 10^{-9})^2}$$

$$= 12.4 \times 10^6 \, N \, C^{-1}$$

These fields can then be added vectorially using geometric methods or by components.

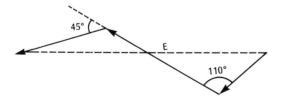

From the length of the resultant vector the field at the sodium ion is about $3.5 \times$ $10^6 \, N \, C^{-1}$. Thus the force on the sodium ion is given by Equation (13-4).

$$F = 3.5 \times 10^6 (1.6 \times 10^{-19})$$

$$= 5.6 \times 10^{-13} \, N$$

Work (W) must be done to move any charged particle into an electric field. Hence, any charged species in an electric field has a potential energy equal to the work done getting it there. Consider for example the work done in bringing a single charge (such as on an ion of charge q_1) into the electric field of a fixed charge, q_2. The little demon in Figure 13-3 would have to do work equal to force times distance or

$$W = -q_1 \int_\infty^r E \, dr \qquad (13\text{-}7)$$

Figure 13.3
Work is done when moving a charge in an electric field

Note that in Equation (13-7) an integration must be performed because the electric field due to the fixed charge gets larger as one gets closer to it. Inserting Equations (13-5) for E and integrating yields

$$W = \frac{kq_2 q_1}{Kr} \qquad (13\text{-}8)$$

This equation is often written

$$W = Vq_1 = \text{Potential Energy of } q_1 \qquad (13\text{-}9)$$

where V, the electric potential, is

$$V = kq_2/Kr \qquad (13\text{-}10)$$

Electric potential has units of joules per

coulomb or volts. It is, therefore, often called voltage. Note that it is a property of the electric field and is independent of the charge (q_1). The electric potential or voltage at a particular point in a network of charged particles is not a vector quantity as was the electric field. It is the sum of all the voltages V_i due to the particles placed at distances, r_i, from the point of interest. Thus

$$V = \sum V_i \qquad (13\text{-}11)$$

Usually absolute voltages as described above are not useful because experimentally only voltage differences can be measured. In neural cells, for example, the voltage difference across the membrane is typically in the neighbourhood of -60 millivolts. That is, the inside of the membrane has an electrical potential which is 60 millivolts lower than the exterior.

Many membranes are permeable to small ions such as chloride (Cl^-) and sodium (Na^+), but are impermeable to charged molecules. In this case, a stable voltage difference across the membrane can be produced simply by a rearrangement of ion concentrations on either side. In order to see how this comes about imagine the circumstances shown in Figure 13-4. Originally the two compartments contained solutions of sodium chloride. A soluble sodium salt, NaQ, was then added to the inside compartment and this salt ionized to produce sodium ion and a large singly charged species, Q^-. A semipermeable membrane separates the two compartments. It allows free passage of Na^+ and Cl^-, but its pores are too small to let the negatively charged large species (Q^-) through. For chloride ion, sodium ion, and the charged species the concentration of electronic charge in moles per litre is the same as their concentration in moles per litre because each carries one electrical charge. Inside, the concentration of charge is $[Cl^-]_i$ moles per litre for Cl^- and $[Na^+]_i$ moles per litre for Na^+ and outside $[Cl^-]_o$ and $[Na^+]_o$ respectively. By using the very simple ideas that like charges repel and unlike charges attract it can be seen that the tendency of the negatively charged species would be to push chloride ions outside and pull sodium ions inside. Of course as soon as this happens a voltage difference develops between the two sides. An equilibrium will be achieved when all forces balance. This equilibrium, called a Donnan equilibrium, is charac-

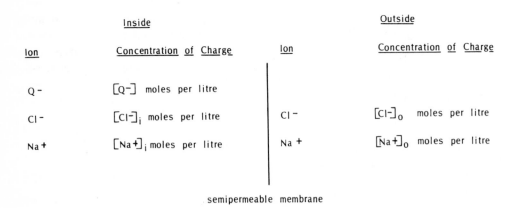

Figure 13.4
Two compartments separated by a semipermeable membrane

teristic to circumstances such as shown in Figure 13-4 and can easily be demonstrated in the laboratory. It is also possible to quantitatively calculate the equilibrium voltage difference, V_{eq}, as well as the equilibrium ion concentrations $[Cl^-]_i$, $[Cl^-]_o$, $[Na^+]_i$ and $[Na^+]_o$.

First it must be realized that the whole system in Figure 13-4 is at thermal equilibrium. When such a situation holds the Boltzmann equation (Chapter 10) may be applied to indicate how the ions should be distributed between the two potential energies (inside and outside). Considering chloride ions first and using Equation (13-9) for the ion potential energies the Boltzmann equation becomes

$$\frac{N}{N'} = e^{-[(V_{outside} - V_{inside})q/kT]}$$

$$= e^{V_{eq}q/kT} \qquad (13\text{-}12)$$

In this equation N and N' correspond to the numbers of chloride ions inside and outside respectively. If these are changed to the molar concentrations $[Cl^-]_i$ and $[Cl^-]_o$ then we must also replace the Boltzmann constant k by the gas constant, R (8.32 J mol^{-1} K^{-1}) and q by its corresponding molar quantity, $n\mathscr{F}$ where $\mathscr{F}$, the Faraday, is the total charge in coulombs of one mole of singly charged ions, and n is the number of charges carried by the ion. $\mathscr{F}$ has a value of 96,500 coulombs per mole. Thus

$$\frac{[Cl^-]_i}{[Cl^-]_o} = e^{V_{eq}n\mathscr{F}/RT} \qquad (13\text{-}13)$$

which can also be written as

$$V_{eq} = \frac{RT}{n\mathscr{F}} \ln\left(\frac{[Cl^-]_i}{[Cl^-]_o}\right) \qquad (13\text{-}14)$$

This equation is known as the Nernst equation. Note that it precisely relates the equilibrium voltage drop across the membrane

(called the membrane potential) to the ion concentrations on either side.

If these same relationships are applied to the sodium ion then one obtains

$$\frac{[Na^+]_i}{[Na^+]_o} = e^{-V_{eq}n\mathscr{F}/RT} \qquad (13\text{-}15)$$

and

$$V_{eq} = -\frac{RT}{n\mathscr{F}} \ln\left(\frac{[Na^+]_i}{[Na^+]_o}\right) \qquad (13\text{-}16)$$

Note also by comparing Equation (13-13) and Equation (13-15) that

$$[Na^+]_i[Cl^-]_i = [Na^+]_o[Cl^-]_o, \qquad (13\text{-}17)$$

an equation which is extremely valuable in finding the equilibrium values of the ion concentrations.

There are two other equations which relate the concentrations of ions at equilibrium for a case such as that of Figure 13-4. First of all, the total ion concentration is constant so that

$$[Cl^-]_i + [Cl^-]_o = \text{constant} \qquad (13\text{-}18)$$

and

$$[Na^+]_i + [Na^+]_o = \text{constant} \qquad (13\text{-}19)$$

Secondly, each side of the system is very nearly neutral. The slight differences in concentration required to establish the membrane potential are negligible. That is, the amount of positive charge must equal the amount of negative charge on each side. Therefore

$$[Q^-]_i + [Cl^-]_i = [Na^+]_i \qquad (13\text{-}20)$$

and

$$[Cl^-]_o = [Na^+]_o \qquad (13\text{-}21)$$

Table 13-3 outlines the important equations which allow one to find the ion concentrations at Donnan equilibrium.

Table 13-3
Equations for Finding Ion Concentrations
at Donnan Equilibrium

(1) Thermal Equilibrium	$[Na^+]_i[Cl^-]_i = [Na^+]_o[Cl^-]_o$
(2) Conservation of ions	$[Cl^-]_i + [Cl^-]_o = $ constant
	$[Na^+]_i + [Na^+]_o = $ constant
(3) Neutrality	$[Q^-]_i + [Cl^-]_i = [Na^+]_i$
	$[Cl^-]_o = [Na^+]_o$

EXAMPLE

Two compartments of a solution are separated by a semipermeable membrane which allows free passage of ions but blocks any transfer of macromolecules. A solution of sodium hyaluronate (which ionizes as shown to produce a negatively charged polyelectrolyte and sodium ions) is placed in the inside compartment and a salt solution is added to both sides. The initial concentrations are shown below. Assume a temperature of 20°C.

Inside		Outside	
Ion	Conc. of Charge	Ion	Conc. of Charge
Q^-	0.075 mol l^{-1}		
Na^+	0.165 mol l^{-1}	Na^+	0.090 mol l^{-1}
Cl^-	0.090 mol l^{-1}	Cl^-	0.090 mol l^{-1}

What will be the ion concentrations in the compartments and the membrane potential when Donnan equilibrium is reached? The equations from Table 13-3 which correspond to this case are

(i) $[Cl^-]_i[Na^+]_i = [Cl^-]_o[Na^+]_o$
(Thermal equilibrium)
(ii) $0.075 + [Cl^-]_i = [Na^+]_i$ (Neutrality)
(iii) $[Cl^-]_o = [Na^+]_o$ (Neutrality)
(iv) $[Na^+]_i + [Na^+]_o = 0.165 + 0.90$
$= 0.233$ (Conservation of ions)
(v) $[Cl^-]_i + [Cl^-]_o = 0.090 + 0.090$
$= 0.180$ (Conservation of ions)

To solve these equations we need to eliminate all but one variable say, for example, $[Na^+]_o$. Begin with equation (v)

$$[Cl^-]_i + [Cl^-]_o = 0.180$$

Using (i) and (iii) this can be converted to

$$\frac{[Cl^-]_o[Na^+]_o}{[Na^+]_i} + [Na^+]_o = 0.180$$

Using (iii) again

$$\frac{[Na^+]_o^2}{[Na^+]_i} + [Na^+]_o = 0.180$$

Using (iv)

$$\frac{[Na^+]_o}{0.255 - [Na^+]_o} + [Na^+]_o = 0.180$$

or

$$[Na^+]_o^2 + 0.255[Na^+]_o - [Na^+]_o^2$$
$$= 0.180(0.255) - 0.180[Na^+]_o$$

Therefore

$$0.435[Na^+]_o = 0.046$$
$$[Na^+]_o = 0.106 \text{ mol } l^{-1}$$

Using (iii)

$$[Cl^-]_o = [Na^+]_o = 0.106 \text{ mol } l^{-1}$$

Using (v)

$$[Cl^-]_i + 0.106 = 0.180$$
$$[Cl^-]_i = 0.074 \text{ mol } l^{-1}$$

Using (iv)

$$[Na^+]_i + 0.106 = 0.255$$
$$[Na^+]_i = 0.149 \text{ mol } l^{-1}$$

The final concentrations are

Inside		Outside	
Ion.	Conc.	Ion	Conc.
Q^-	0.075 mol l^{-1}		
Na^+	0.149 mol l^{-1}	Na^+	0.106 mol l^{-1}
Cl^-	0.074 mol l^{-1}	Cl^-	0.106 mol l^{-1}

The membrane potential can be found using either the chloride ions or the sodium ions. Thus from Equation (13-14)

$$V_{eq} = \frac{RT}{n\mathscr{F}} \ln\left(\frac{[Cl^-]_i}{[Cl^-]_o}\right)$$

$$= \frac{(8.32)(293)}{1 \times 96,500} \ln\left(\frac{0.074}{0.106}\right)$$

$$= -25 \times 10^{-3} \ln 0.70$$

$$= -9.0 \times 10^{-3} \text{ volts}$$

$$= -9.0 \text{ millivolts}$$

From Equation (13-16)

$$V_{eq} = -\frac{RT}{n\mathscr{F}} \ln\left(\frac{[Na^+]_i}{[Na^+]_o}\right)$$

$$= -25 \times 10^{-3} \ln\left(\frac{0.149}{0.106}\right)$$

$$= -9.0 \text{ millivolts}$$

The property of the membrane which allows a Donnan-type equilibrium to be established is that it contain pores which are large enough to allow the passage of ions, but too small to allow polyelectrolytes through. There is no requirement that the membrane be of biological origin and many synthetic membranes work just fine. Sometimes, when the polyelectrolyte concentration is very high it is necessary to make corrections to membrane potentials due to osmotic effects.

Over the years the development and improvement of microelectrodes and sensitive electronic equipment has made the measurement of membrane potentials a reasonably straightforward procedure. It is possible to measure membrane potentials in live axons and watch how these potentials change as the neural spike passes. Normally, if the neural cell is alive, but not stimulated (i.e. producing neural spikes) the axon membrane has a "resting potential" of about −60 mV. Other types of cells, such as muscle cells, also have a characteristic resting potential. Using Equations (13-14) and (13-16) it is possible to estimate what the concentration ratios (inside/outside) would be for ions such as Na^+, Cl^- and K^+ (potassium ion) if the axon membrane behaved like the one in the previous example. Table 13-4 shows the results of such calculations and the concentration ratios found experimentally for squid axon.

Table 13-4
Concentration Ratios (Calculated Assuming Donnan Equilibrium and Experimentally Measured in Squid Axon)

Ion Ratio	Calculated Value using Equations (13-14) & (13-16)	Experimental Ratios
$[Cl^-]_i/[Cl^-]_o$	9.1×10^{-2}	10×10^{-2}
$[K^+]_i/[K^+]_o$	11	20
$[Na^+]_i/[Na^+]_o$	11	0.11

The concentration ratios calculated and measured for chloride ion are very close, indicating that this species is, indeed, very near equilibrium.

For potassium and especially sodium there is no agreement at all between the calculated and measured ratios. There is somewhat more potassium inside than expected and about one hundred times less sodium inside than expected. The system is in a much higher energy state than would be expected if thermal processes alone were involved. Thus, as far as these two ions are concerned, the membrane system is very far from Donnan equilibrium. Using Equation (13-16) one would expect that the measured values of the sodium and potassium concentration ratios should yield membrane potentials of about $+55\,\mathrm{mV}$ and $-75\,\mathrm{mV}$ respectively. Using Equation (13-14), the corresponding potential for chloride ion is $-58\,\mathrm{mV}$. These values, along with the resting potential, are shown in Figure 13-5.

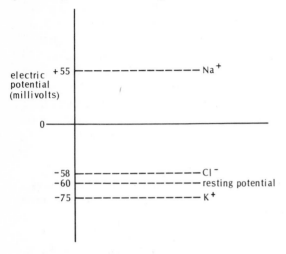

Figure 13.5
The resting potential and expected potentials for chloride, sodium and potassium ions

In order to maintain the sodium ion and potassium ion concentration ratios described earlier, it is necessary for these ions to be actively pumped through the membrane. In the case of sodium ions, not only do they have to be transferred from inside to outside, but this has to be done against an electric potential difference and against a huge sodium ion concentration gradient. This process requires a sizeable amount of energy so that the membrane has to have all the machinery for converting chemical energy (usually from the molecule adenosine triphosphate) into a form usable by the sodium pumps. Thus the resting membrane is in a state of dynamic equilibrium, dependent on the energy supply available. The only time a biological membrane such as this reaches Donnan equilibrium is after death.

The dynamic equilibrium associated with an axon membrane at resting potential is designed to maintain the ion concentration ratios at the values listed in Table 13-4. For this to happen the net ionic current density, I_d, through the membrane must be zero. Ohm's law states that current is related to potential difference and conductance by the equation

$$I = gV \qquad (13\text{-}22)$$

If membrane conductivity is used then this equation becomes

$$I_d = g_m V \qquad (13\text{-}23)$$

where I_d is the current density described earlier. For the resting membrane

$$I_d = 0 = (g_{\mathrm{Na}^+})(V_r - V_{\mathrm{Na}^+}) + (g_{\mathrm{K}^+})(V_r - V_{\mathrm{K}^+})$$
$$+ (g_{\mathrm{Cl}^-})(V_r - V_{\mathrm{Cl}^-})$$
$$+ \text{terms due to other ions} \qquad (13\text{-}24)$$

Since the terms due to other ions are fairly small it is possible to calculate the resting membrane potential, V_r, using the membrane conductivity values from Table 13-1 and the ionic potentials from Figure 13-5. The resulting value (about $-65\,\mathrm{mV}$) compares favorably with the experimental values of about $-60\,\mathrm{mV}$ mentioned earlier especially in view of the relative uncertainties present in the values for the membrane conductivities and the ionic potentials. It is clear that the axon membrane is a finely tuned system. The conductances are precisely determined for each ion to give the appropriate resting potential values. The pumps are set to maintain the ion concentration ratios (inside/outside) at precisely the right level by pumping excess ion which leaks through right back again. It is felt in some quarters that the operation of the sodium pump controls potassium and chloride ion ratios by default although this has not been clearly established at this time. The detailed operation of the ion pumps is not well understood at all.

13-3 Propagation of the Neural Spike

When an electric potential of about 20 mV is applied to a small section of axon membrane at resting potential a momentary cataclysmic event takes place. It's almost as if the membrane temporarily forgets it's a membrane and becomes a sieve instead. The sodium and potassium ions, which had previously been held at the concentration ratios given in Table 13-4 suddenly strive to achieve Donnan equilibrium values. The result is the neural spike which is shown in Figure 13-6 as a plot of membrane potential against time in milliseconds. The spike is a relatively strong (about +40 mV), but short-lived (about 1 millisecond) positive potential which moves rapidly down the axon. Just prior to the spike there is a shoulder, the pre-spike potential, which is the response due to the applied 20 mV potential. Notice that the pre-spike potential increases to a potential about fifteen

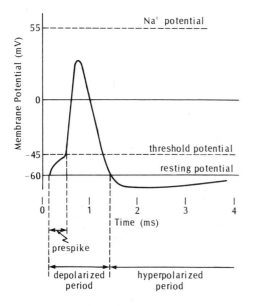

Figure 13.6

The neural spike

millivolts above resting potential. At this point, the threshold potential, the membrane conductance properties suddenly changes and the positive spike develops. The exact structural or functional mechanism behind this change is not known but is the subject of a great deal of research.

During the pre-spike potential and the spike itself the axon membrane is said to be depolarized because the potential is above the resting level. Just following the spike the membrane becomes hyperpolarized (more negative than resting potential) for a few milliseconds. During the first part of this period, the so-called absolute refractory period, a second spike cannot be initiated at all. Later in the hyperpolarized period it is possible but more difficult to initiate a second spike because a greater pre-spike potential has to be supplied to raise the potential to the threshold level. The hyperpolarization period is, therefore, a kind of busy signal or dead time.

A better understanding of the neural spike can be obtained if the sodium and potassium conductivities are followed during the event. These quantities are shown in Figure 13-7. The change in sodium conductivity is incredible, from a value of $0.01 \times 10^{-3} \, \Omega^{-1} \, cm^{-2}$ prior to threshold to a peak value of about $30 \times 10^{-3} \, \Omega^{-1} \, cm^{-2}$, a factor of about a thousand. The change in potassium conductivity is not so spectacular ($0.3 \times 10^{-3} \, \Omega^{-1} \, cm^{-2}$ to about $10 \times 10^{-3} \, \Omega^{-1} \, cm^{-2}$) but sizeable nevertheless. The huge increase in sodium conductivity causes a massive current of sodium ions (at least at the microscopic level) and this current of positive charge causes the inside of the axon to become positively charged with respect to the outside. Effectively the sodium ions are striving to reach their equilibrium potential of +55 mV. They never quite make it, however, because (a) their local concentration is depleted and (b) the sodium conductivity rapidly returns to its resting value. The

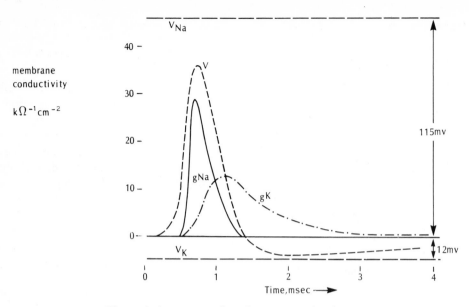

Theoretical reconstruction of a propagated action potential (curve V) and sodium and potassium conductances, using experimental constants appropriate to 18.5°C. (from A. L. Hodgkin and A. F. Huxley, J. Physiol. **117**, 500 (1952).

Figure 13.7

Sodium and potassium membrane conductivities during the neural spike

spike is positive for only a very brief period because potassium ions are flowing out in an attempt to reach their own equilibrium potential of −75 millivolts. This outgoing positive current balances the earlier incoming current and the membrane quickly returns to the resting potential. Additional out-flux of potassium continues for several milliseconds and leads to the hyperpolarized refractory period. In about 3 milliseconds the membrane has returned to resting potential and is ready to fire again. Because of the dead-time associated with the refractory period the maximum firing rate is about 300 spikes per second.

The neural spike which has just been described travels down an axon by a mechanism which has some similarities to an electrical pulse travelling down a cable. In contrast to the cable situation though the neural spike never loses its positive amplitude of +55 mV, regardless of the length of the axon. An ordinary electrical pulse in a cable would gradually lose its amplitude—the longer the cable, the greater the reduction in amplitude. However, a combination of cable theories with the special properties of the axon membrane does provide a reasonable mechanism for spike propagation. First it must be recognized that at any instant the neural spike has depolarized only a very small fraction of the total axon length. The depolarized region is shown schematically and highly magnified in Figure 13-8.

Inside the spike region the membrane po-

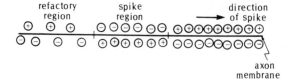

Figure 13.8

Membrane depolarization at the neutral spike

tential is positive, reaching a maximum value of about 40 mV. This potential can be represented by the symbol V_o^*. Outside the spike region the spike potential gradually decreases as would be expected in any electrical cable. The value of the spike potential at some distance away from the spike region can be represented by V^*. The electric field associated with the spike potential exerts a force which tends to move the ions located nearby. Longitudinal internal and external ion currents as well as membrane currents result as shown in Figure 13-9.

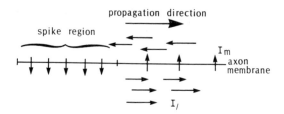

Figure 13.9

Ion currents due to the spike potential

The current densities which are of specific importance are the internal longitudinal current density, I_l, down the axon and small leakage current density through the membrane outside the spike region (I_m). In order to evaluate the importance of these currents in the propagation of the spike it is necessary to consider in greater detail the concepts related to conductivity and resistivity in the axon. Since resistivity is just the inverse of conductivity any comments regarding one will also

apply to the other. There are two resistivities associated with the axon which determine the current densities I_m and I_l. They are respectively the membrane resistivity and the longitudinal core resistivity of the interior of the axon. The membrane resistivity is just the inverse of the membrane conductivity which has already been discussed. Therefore

$$R_m (\Omega \text{ cm}^2) = \frac{1}{g_m (\Omega^{-1} \text{ cm}^{-2})} \quad (13\text{-}25)$$

Below threshold the values of R_m for various ionic species are the inverse of the corresponding g_m values in Table 13-1. The resistance (R) of any particular area (A) of the membrane may be obtained from the relationship

$$R = \frac{R_m}{A} \quad (13\text{-}26)$$

The resistivity (R_l) associated with the interior of the axon (the axoplasm) is a somewhat different quantity from R_m because it is a volume property rather than a surface property. The resistance of a specified volume of axoplasm is given by

$$R = \frac{R_l L}{A} \quad (13\text{-}27)$$

where in this case A is the cross-sectional area of the volume and L is the length of the volume.

EXAMPLE

The resistivity of the squid axon axoplasm is about 30 Ω cm. Find the resistance of an axon 10 cm long having a radius of 0.025 cm.

From Equation (13-27)

$$R = 30 \left(\frac{10}{\pi (0.025)^2} \right)$$

$$= 1.5 \times 10^5 \text{ ohm}$$

It is sometimes convenient to express axon resistance on a per unit length basis such as per centimetre of length. Since the axon is cylindrically shaped then from Equation (13-26)

Membrane resistance for 1 cm of axon

$$= \frac{R_m}{2\pi r(1)} \qquad (13\text{-}28)$$

and from Equation (13-27)

Longitudinal resistance per unit length

$$= \frac{R_l(1)}{\pi r^2} \qquad (13\text{-}29)$$

where r is the radius of the axon. It is these quantities which are related to the membrane and longitudinal current densities.

From Ohm's law, the magnitude of I_l depends on the voltage difference or gradient, dV^*/dx (the rate of change of V^* with distance, x, from the spike region) along the axon and the actual resistance per unit length of the axoplasm ($R_l/\pi r^2$). Thus

$$I_l = -\frac{\pi r^2}{R_l}\left(\frac{dV^*}{dx}\right) \qquad (13\text{-}30)$$

The negative sign in Eq. (13-30) arises because current is conventionally considered to flow from more positive to less positive potentials. If the membrane was a perfect insulator and no leakage current of ions occurred then I_l would be a constant. Below threshold, membranes do have some small conductivity values, however, and I_l decreases with distance from the spike because of this leakage. The value of the membrane current density is equal to the rate at which the longitudinal current density decreases along the axon:

$$I_m = -\frac{dI_l}{dx} \qquad (13\text{-}31)$$

In other words the longitudinal current which is lost appears as membrane current. In addition, from Ohm's law it is also true that I_m at a point on the membrane is given by

$$I_m = \frac{V^*}{R_m/2\pi r} \qquad (13\text{-}32)$$

where V^* is the potential difference at that point due to the spike potential and $R_m/2\pi r$ is the membrane resistance of a unit length. Note that this equation applies only to regions of the membrane where the potential is below threshold.

From Equations (13-31), (13-32) and the differentiation of Equation (13-30)

$$-\frac{dI_l}{dx} = \frac{V^*}{R_m/2\pi r}$$

$$= \frac{\pi r^2}{R_l}\frac{d^2 V^*}{dx^2}$$

and

$$V^* = \frac{r}{2}\frac{R_m}{R_l}\frac{d^2 V^*}{dx^2} \qquad (13\text{-}33)$$

The solution to this differential equation is

$$V^* = A\,\exp\left[-x\left(\frac{2R_l}{rR_m}\right)^{1/2}\right]$$
$$+ B\,\exp\left[+x\left(\frac{2R_l}{rR_m}\right)^{1/2}\right] \qquad (13\text{-}34)$$

However V^* must approach zero at distances, x, which are very large and hence the constant, B, must be zero. When x is zero V^* must equal the spike potential which means that A is equal to V_o^*. Thus

$$V^* = V_o^*\,\exp\left[-x\left(\frac{2R_l}{rR_m}\right)^{1/2}\right] \qquad (13\text{-}35)$$

and V^* is seen to fall off exponentially with distance from the spike region. Notice that V^* decreases to $1/e$ of its original value when x has a magnitude of $(rR_m/2R_l)^{1/2}$. This distance is a characteristic of all electrical cables and is called the length constant. The exponential character of V^* is shown in Figure 13-10 for squid axons of two different radii.

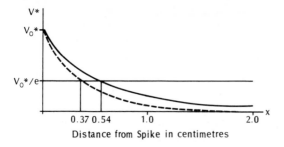

Figure 13.10

The exponential character V^* in squid axon: solid line $r = 0.025$ cm; broken line $r = 0.012$ cm. The corresponding length constants are indicated

---

EXAMPLE

For the squid axon the membrane resistivity $R_m = 700\ \Omega\ cm^2$ and the longitudinal core resistivity $R_l = 30\ \Omega\ cm$. Typically the squid axon has a radius of 0.25 mm or 0.025 cm. Find (a) the length constant associated with the axon and (b) the distance from the spike region at which the potential due to the spike reaches +15 mV. Assume V_o^* is +55 mV.

(a) The length constant $= \left(\dfrac{rR_m}{2R_l}\right)^{1/2}$

$= \left(\dfrac{0.025 \times 700}{2 \times 30}\right)^{1/2}$

$= 0.54$ cm

(b) Using Equation (13-33) and results from (a)

$$15 = 55e^{-x/0.77}$$

$$ln\left(\frac{15}{55}\right) = \frac{-x}{0.54}$$

$$x = 0.70\ cm$$

The example above demonstrates that the positive potential due to the spike extends a sizeable distance down the axon from the spike region into a region which is still at resting potential. This positive potential acts exactly as the earlier applied potential and produces a pre-spike potential in the new region. If the value of V^* is greater than the potential required to exceed the threshold potential then a new spike is produced and a new spike region develops downstream from the first one. The same ion movements occur and a new spike of the same amplitude as the first one is generated. This new spike will in turn lead to the production of a new one still further downstream and the process continues down the full length of the axon. Spike propagation thus takes place with no loss of amplitude in the transmission process.

The mechanism of spike progagation is therefore closely involved with the dynamic equilibrium which characterizes the resting membrane, and with the physics contained in Equation (13-35). The length constant, which comes from this equation can also be used to estimate the propagation velocity of the spike as it moves down the axon. Figure 13-10 also demonstrates the length constants of two axons differing only in their radii. The axon with the larger radius has a larger length constant and the magnitude of the potential V^* falls off more gradually with distance down the axon. In this axon, the first spike can initiate a new one in a more distant region of the fibre than is possible in the smaller diameter axon. Therefore, in larger axons the spike propagates with bigger steps and travels faster. The velocity of spike propagation is thus proportional to the length constant:

$$v \propto \sqrt{\frac{rR_m}{2R_l}} \qquad (13\text{-}36)$$

For axons having the same values of R_m and R_l Equation (13-36) can be rearranged to yield:

$$\frac{v^2}{r} = constant \qquad (13\text{-}37)$$

or

$$\frac{v_1^2}{r_1} = \frac{v_2^2}{r_2} \qquad (13\text{-}38)$$

Equation (13-38) is useful for comparing the propagation velocity of spikes in similar axons with different radii.

EXAMPLE

Two axons (sciatic nerve) from the frog have radii of 20×10^{-6} m and 10×10^{-6} m. In the first axon a spike was timed electronically to travel a distance of 0.02 m in 5×10^{-4} s. Estimate the time required for a spike to travel a similar distance in the second axon.

$$\text{Velocity} = \frac{\text{distance}}{\text{time}}$$

$$= \frac{0.02}{5 \times 10^{-4}}$$

$$= 40 \text{ m s}^{-1}$$

From Equation (13-38)

$$\frac{(40)^2}{20 \times 10^{-6}} = \frac{(v_2)^2}{10 \times 10^{-6}}$$

$$v_2^2 = 800$$

$$v_2 = 28 \text{ m s}^{-1}$$

It is of significant benefit to larger animals to have fast reaction times. This can be achieved by the system evolving axons which have large radii. Such axons would take up more space, however, and space is at a premium. The significance of this becomes apparent if, for example, one considers the optic nerve bundle of the eye, which contains about 10^6 axon fibres. If each one had a radius of 0.05 cm, the radius of the squid axon, then the total cross-sectional area of the optic nerve bundle of the eye would have to be about 8000 cm². This is many times larger than the cross-section of the eye itself.

More advanced species such as mammals have reduced the axon radius and increased R_m extensively. This is the role of myelination which we described at the beginning of the chapter. The neural spikes in these systems appear only at the gaps in the myelin, the so-called nodes of Ranvier. The spike at one node produces a potential which causes the pre-spike potential at the next. Thus in these systems the spike "hops" from node to node and the transmission speed is, therefore, much faster. As a result, the myelinated nervous cells of these animals can carry a great deal more information and with extremely high efficiency. This beneficial effect of increasing R_m is also apparent from Equation (13-36).

There are many other fascinating topics in the biophysics of neural systems which cannot be covered here. It is a rapidly growing field, spurred on by the possibility that eventually we may be able to understand how the brain, a highly complex neural system, functions. For the interested student we suggest such books as *Nerve, Muscle and Synapse* by Bernard Katz and *Membranes, Ions and Impulses* by K. C. Cole for further reading.

Problems

1. In a particular cell Na^+ ions are pumped from the inside to the outside at a rate of 10^5 s^{-1}. Calculate the corresponding electrical current in amperes.

2. In 10 seconds 8×10^5 Na^+ ions pass through a section of membrane having a total area of 50 μm^2. What is (a) the electrical current and (b) the current density?

3. In an action potential about 4.3×10^{-12} mol cm^{-2} of sodium ions enter the axon in about one millisecond. Find the current density associated with this flow of ions.

What is the electrical current if the action potential involves a membrane area of $5 \ \mu m^2$?

4. Two negative charges of 10^{-6} C each are placed in water and separated by a distance of 0.10 m. A positive charge of 10^{-8} C is placed exactly midway between the two negative charges.
   (a) What is the electric field, force and electrical potential experienced by the positive charge?
   (b) What is the electric field, force and electrical potential experienced by either negative charge?

5. Find the force on the chloride ion (Cl_1^-) in the diagram below. All the ions lie in the same plane and the solvent is water.

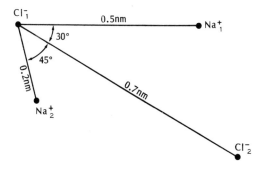

6. A membrane which is permeable to Na^+ and Cl^- separates two solutions, one of which contains a large anion Q^-, to which the membrane is impermeable. The initial concentrations are (using i and o subscripts to designate the compartments):

$$[Na^+]_o = [Cl^-]_o = 0.20 \ mol \ l^{-1}$$
$$[Na^+]_i = 0.50 \ mol \ l^{-1}$$
$$[Cl^-]_i = 0.20 \ mol \ l^{-1}$$
$$[Q^-]_i = 0.30 \ mol \ l^{-1}$$

If the temperature is 20°C, what potential difference develops across the membrane?

7. A membrane which is permeable to small ions, (Na^+, Cl^-) but impermeable to the large ions (Q^-) separates two compartments (i and o). The system is at Donnan equilibrium. The concentrations $[Q^-]_o$ is zero, $[Na^+]_i$ is 0.36 mol l^{-1} and $[Cl^-]_o$ is 0.14 mol l^{-1}. Find $[Na^+]_o$, $[Cl^-]_i$ and $[Q^-]_i$ and the membrane potential.

8. In the giant axon of the squid, the measured Na^+, K^+ and Cl^- concentrations in millimoles per litre are:

	Outside	Inside
Na^+	150	15
K^+	5	150
Cl^-	125	10

What are the Cl, Na and K equilibrium potentials at 20°C?

9. Using Equation (13-24) and the necessary data from Table 13-1 and Figure 13-5, show that the resting potential of the squid axon is about −65 mV.

10. The resistivity of a cellular fluid is $60 \ \Omega$ cm. What is the conductance of a tubule of diameter 10 μm and length 100 μm, filled with this fluid?

11. For the purpose of calculation of electrical current through the human body, the body fluid can be considered as a solution of resistivity 65 Ω cm at normal body temperature. Calculate the resistance of one arm, considering it to be 75 cm long and of a uniform cross-sectional area of radius 5 cm. Similarly, calculate the resistance of one finger, considering it to be 10 cm long and of 4 cm^2 cross-section. The actual body resistance depends upon the magnitude and the direction of the current and on a number of other factors, such as the time of the day, etc.

12. An axon from a lobster has a diameter of 70 μm. The membrane resistivity is $2000 \ \Omega \ cm^2$ and the resistivity of the

axoplasm is $60\,\Omega$ cm. Assume that the neural spike reaches a maximum positive potential of $+30$ mV.

(a) What is the length constant associated with this axon?

(b) At what distance from the positive spike region will the positive potential drop to $+15$ mV?

13. In a certain crab nerve of radius 15 μm a positive spike potential of $+29$ mV is observed. If the threshold potential of this nerve is 18 mV above the resting potential, what is the furthest distance down the axon that a second spike can be triggered from the first one?

(For crab nerve $R_m = 5{,}000\,\Omega$ cm^2 and $R_l = 60\,\Omega$ cm).

(The potential at any point on the resting membrane is $V_r + V^*$).

14. Two axons, A and B, from the same species have identical membrane and axoplasm resistivities. However, the core resistance of a one centimetre length of axon A is only 1/4 the core resistance of axon B. Calculate the expected ratio of the velocities of propagation in these two axons.

Appendix 1
Mathematical Skills

A1-1 Introduction

A knowledge of the following mathematical topics is useful for reading this text.

1. Simple algebra
2. Exponentials and logarithms
3. Simple differentiation and integration
4. Graphing of simple functions
5. Sine and cosine functions

Most students will already be familiar with these topics, but will require a brief review. This Appendix attempts to fill that need. If, at the end of the exercises, a student feels particularly weak in a mathematical skill, he should refer to a mathematical text.

If we tackle a typical biological problem on, for example, exponential growth and decay, the reasons for knowing these mathematical operations will quickly become apparent.

TYPICAL PROBLEM: EXPONENTIAL GROWTH AND DECAY

Let us consider a microbiologist who is studying some of the characteristics of a certain strain of bacteria. For a particular experiment he requires 10^6 cells in a test-tube, but at time $t = 0$ he has only 100 cells. Under optimum conditions the growth rate constant of his bacteria is $3.47 \times 10^{-4}\,s^{-1}$ so he can calculate the length of time it takes to obtain 10^6 cells.

We will show this calculation in a series of steps and then amplify each step later.

STEP 1 *Statement of the Problem in Mathematical Terms*

Almost every living species has the same type of growth equation. In words "the rate of change of a population at some time, t, is proportional to the population itself at that time." This is very obvious for humans in that the more people there are, the more children will be born over a certain time interval.

If we designate N as being the number of cells, then dN/dt is the rate of change of the bacterial population with respect to time.

From the preceding argument:

$$\frac{dN}{dt} \propto N \quad (\propto \equiv \text{is proportional to})$$

or

$$\frac{dN}{dt} = kN \text{ where } k \text{ is a constant which}$$
we will call the "growth rate constant"

STEP 2 *Putting the Equation in a Useable Form*

We must integrate the preceding equation in order to put it into a form we can use.

We will begin by rearranging the equation to yield

$$\frac{dN}{N} = k\,dt$$

195

If we require that the number of bacteria be N_o when $t = 0$ and N when $t = t$, then

$$\int_{N=N_o}^{N=N} \frac{dN}{N} = k \int_{t=0}^{t=t} dt$$

Integrating yields (integral of $1/N$ is $ln\ N$)

$$\left. ln\ N\ \right]_{N=N_o}^{N=N} = \left. kt\ \right]_{t=0}^{t=t}$$

$$ln\ N - ln\ N_o = k(t) - k(0)$$

or

$$\boxed{ln\ \frac{N}{N_o} = kt} \qquad (A1\text{-}1)$$

We now have the equation in a useable form.

STEP 3   *Applying the conditions and solving*

In our particular problem $N = 10^6$, $N_o = 10^2$ and $k = 3.47 \times 10^{-4}$ s^{-1}. Thus:

$$ln\ \frac{10^6}{10^2} = 3.47 \times 10^{-4}t$$

$$ln\ 10^4 = 3.47 \times 10^{-4}t$$

But $ln\ 10^4 = 9.210$

$$t = \frac{9.210}{3.47 \times 10^{-4}}$$

$$= 2.66 \times 10^4\ s$$

$$= 7.37\ hr.$$

Amplification of Step 1.

Many people experience serious difficulties when they have to put their observations on a mathematical basis. Most scientists do it by varying one parameter at a time and then drawing graphs.

Thus, if the microbiologist plotted the rate of change of the bacterial population with respect to time, dN/dt, against N he would have obtained a graph like the following

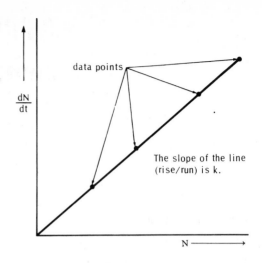

data points

The slope of the line (rise/run) is k.

Thus $dN/dt = kN$ is an equation of the form $y = ax$ where y is the dependent variable, a is a constant, x is the independent variable.

You should be familiar with the graphs of several other simple functions, such as

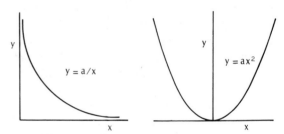

$y = a/x$

$y = ax^2$

Section A1-3 provides exercises on simple graphing.

Amplification of Step 2.

Essentially all that is needed here is a brief review of the concepts of differential and integral calculus. Let's begin with differentiation.

If we have, as an example, the equation $y = kx^2$, "k" being a constant, we can say that y is a function of x or $y = f(x)$. The basic definition of the derivative of y with respect to x (written dy/dx) is

$$\frac{dy}{dx} = \lim_{\Delta x \to 0} \frac{f(x + \Delta x) - f(x)}{\Delta x}$$

where $\lim_{\Delta x \to 0}$ means the limit as Δx approaches zero.

Thus if

$$y = kx^2$$

$$\frac{dy}{dx} = \lim_{\Delta x \to 0} \frac{k(x + \Delta x)^2 - k(x^2)}{\Delta x}$$

$$= \lim_{\Delta x \to 0} \frac{k(x^2 + 2\Delta x \cdot x + \Delta x^2) - kx^2}{\Delta x}$$

$$= \lim_{\Delta x \to 0} \frac{kx^2 + 2kx\Delta x + k\Delta x^2 - kx^2}{\Delta x}$$

$$= \lim_{\Delta x \to 0} \frac{2xk\Delta x + k\Delta x^2}{\Delta x}$$

$$= \lim_{\Delta x \to 0} 2kx + k\Delta x$$

$$= 2kx$$

Thus if

$$y = kx^2$$

$$\frac{dy}{dx} = 2kx$$

Thus $2kx$ is the derivative of kx^2 with respect to x. For all functions of the form

$$y = kx^n$$

$$\boxed{\frac{dy}{dx} = nkx^{n-1}} \qquad \text{(A1-2)}$$

The biggest hangups for most people in the above example are:

1. difficulty in comprehending a "limit"
2. dividing by Δx even though Δx is approaching zero

If you are one of these people, the following example might help. The equation $s = \frac{1}{2}gt^2$ is the familiar equation of free fall, s is the distance travelled, g is a constant (9.8 m s^{-2}) and t is time.

If we plot a graph of s vs. t it would look like:

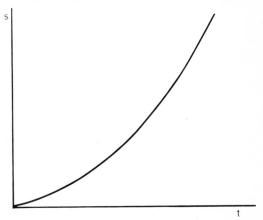

Suppose some sky-diver wants to know his velocity 3 seconds after he jumps from the plane. If he was a really smart sky-diver he probably wouldn't jump at all. However, if he was only slightly smart he'd realize that

$$\text{velocity} = \frac{ds}{dt} = \frac{d(1/2\,gt^2)}{dt} = gt$$

so that at 3 seconds his downward velocity would be $9.8 \times 3 = 29.4 \text{ m s}^{-1}$

Let's assume, however, that this particular sky-diver is blissfully ignorant of differential calculus. How would he find his downward velocity at 3 seconds? Probably he'd begin by finding the distance travelled after 4 seconds and comparing it to the distance travelled after 3 seconds. He is sufficiently worldly to realize that $\text{velocity} = \dfrac{\text{distance travelled}}{\text{time taken}}$. His estimate of his velocity at three seconds would then be

$$\frac{\left(\begin{array}{c}\text{distance} \\ \text{travelled in 4 s}\end{array}\right) - \left(\begin{array}{c}\text{distance} \\ \text{travelled in 3 s}\end{array}\right)}{(4 - 3)}$$

$$= \frac{\frac{1}{2}(9.8)(4)^2 - \frac{1}{2}(9.8)(3)^2}{1}$$

$$= 4.9(16-9)$$
$$= 4.9(7)$$
$$= 34.3 \text{ m s}^{-1}$$

But when he hits the ground, a buddy tells him that he would have got a better estimate if he had compared distance travelled in $3\frac{1}{2}$ and 3 seconds. So he flies up and jumps again. This time he gets a velocity estimate of

$$\frac{4.9 \times (3.5)^2 - 4.9 \times (3.0)^2}{(3.5-3.0)}$$

$$= \frac{4.9[(3.5)^2 - (3.0)^2]}{(3.5-3.0)}$$

$$= 31.8 \text{ m s}^{-1}$$

Again when he hits the ground his buddy says that there is still room for improvement and suggest that he compare the distance covered in 3.1 and 3.0 seconds. So up he goes. This time he gets

$$\frac{4.9 \times (3.1)^2 - 4.9 \times (3.0)^2}{(3.1-3.0)}$$

$$= \frac{4.9[(3.1)^2 - (3.0)^2]}{(3.1-3.0)}$$

$$= 29.9 \text{ m s}^{-1}$$

He keeps trying. Next time he compares 3.01 and 3.00 seconds. He obtains

$$\frac{4.9(3.01)^2 - 4.9(3.00)^2}{3.01-3.00}$$

$$= 29.5 \text{ m s}^{-1}$$

Then he compares 3.001 and 3.000 seconds and obtains

$$\frac{4.9(3.001)^2 - 4.9(3.000)^2}{3.001-3.000}$$

$$= 29.45 \text{ m s}^{-1}$$

While at this point the sky-diver collapses from exhaustion, it's easy to see the trend he is establishing. The number he appears to be approaching as he chooses smaller and smaller

intervals away from 3 seconds is 29.4 m s^{-1}. This sky-diver was actually going through the process of defining a derivative.

$$\text{i.e. } \lim_{\Delta t \to 0} \frac{f(t+\Delta t) - f(t)}{\Delta t} \qquad \text{(A1-3)}$$

The idea of a limit is simply choosing smaller and smaller Δt's until Δt is essentially approaching zero. The fact that we are dividing by such a small number no longer bothers us because we see from the above trend that the numerator is shrinking along with the denominator.

Almost every manipulation in mathematics is reversible and differentiation is no exception. Integration is simply the reverse of differentiation and is sometimes called anti-differentiation.

Consider, as an example, a car which is uniformly accelerating so that its velocity is increasing linearly (i.e. $v = at$ where $v =$ velocity, $a =$ acceleration, and $t =$ time). If we have the information given by the line in the graph and want to know the distance travelled after a certain time, t, then we must use the process of integration. We know already that velocity $= ds/dt$ so that

$$\frac{ds}{dt} = at$$

A graph of velocity as a function of time is shown in the following diagram.

This can be arranged to give $ds = at\, dt$. This equation says that a "little bit" of the distance travelled (ds) is equal to $at\, dt$. But you can see from the graph that $at\, dt$ is simply the area of one of the slices corresponding to the miniscule time interval dt. But the total distance travelled (s) must be the sum of all the "little bits" or the sum of all the ds's. This is written as

$$s = \int ds$$

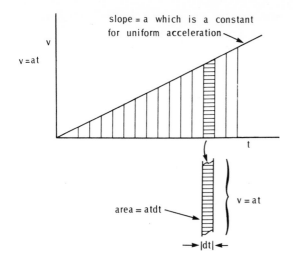

$$y = \frac{3ax^3}{3} = ax^3$$

Unless we add a constant "C", there's no way that ax^3 could equal $ax^3 + 7$. However, if we add C, then

$$ax^3 + C = ax^3 + 7 \text{ and } C = 7$$

Before we leave the subject of integration we should give a little thought to the "definite integral." So far we've seen that the integral of a function corresponds to the area under a curve obtained when the function is graphed. In the case of a definite integral, we have definite values of x between which we must find the area.

If, in our example, we consider the problem of the car experiencing uniform acceleration, we can find the distance travelled between 4 and 6 seconds after it starts. This is done simply by applying limits of integration. Say the acceleration "a" is 10 m s^{-2}.

$$s = \int_4^6 10t \, dt = \frac{1}{2}10t^2 \bigg]_4^6$$

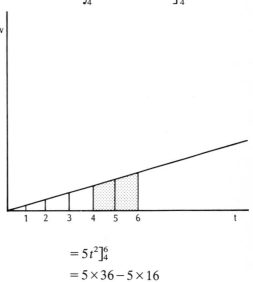

$$= 5t^2 \big]_4^6$$
$$= 5 \times 36 - 5 \times 16$$
$$= 180 - 80$$
$$= 100 \text{ m}$$

where the integral sign ∫ means "the sum of." Thus, $s = \int ds =$ sum of the areas of all the slices $= \int at \, dt$. But the sum of the areas of all the slices is just the area of the triangle under the line of the graph. This is one half of the area of a rectangle, whose length is t and whose height is at, or $\frac{1}{2}at^2$. Thus, $s = \frac{1}{2}at^2$. We have shown, therefore, that $\int at \, dt = \frac{1}{2}at^2$. In general if $f(x) = ax^n$

$$\boxed{\int f(x) \, dx = \frac{a}{n+1} x^{n+1} + C \text{ for } (n \neq -1)}$$

(A1-4)

The constant C arises for the following reason. Consider the function

$$y = ax^3 + 7$$

Clearly

$$\frac{dy}{dx} = 3ax^2$$

If we integrate this we should return to the original equation

$$\int dy = \int 3ax^2 \, dx$$

By using limits of integration we have found the area of the region under the curve bounded by $t = 4$ s and $t = 6$ s.

Amplification of Step 3

The greatest difficulties in this step are the logarithms and exponentials. Logarithms and exponentials are opposite functions in the same sense that integration and differentiation are opposite processes. That is $\int f(x)\, dx = g(x)$ implies that $dg(x)/dx = f(x)$ and $\log_e x = y$ implies that $e^y = x$.

The number "e" has a value of $2.71828\ldots$ and has a very complicated history and some amazing characteristics.

Some early mathematician found that the function e^x was equal to an infinite but converging series

$$e^x = 1 + x + \frac{x^2}{1\cdot2} + \frac{x^3}{1\cdot2\cdot3}$$

$$+ \frac{x^4}{4!} + \frac{x^5}{5!} + \ldots \quad \text{(A1-5)}$$

where 4! means $4\times3\times2\times1$

putting $x = 1$ in the above and adding up several terms will convince you that $e = 2.71828\ldots$

The amazing thing about e^x is that when it is differentiated or integrated, the result is still e^x. This is easy to prove.

$$\frac{de^x}{dx} = \frac{d}{dx}\left(1 + x + \frac{x^2}{1\cdot2} + \frac{x^3}{1\cdot2\cdot3}\right.$$

$$\left. + \frac{x^4}{4!}\ldots\ldots\right)$$

$$= 0 + 1 + x + \frac{x^2}{1\cdot2} + \frac{x^3}{1\cdot2\cdot3} + \ldots\ldots$$

$$= e^x \quad \text{(A1-6)}$$

It is also important that you know that

$$\boxed{\frac{d}{dx}(\ln x) = 1/x} \quad \text{(A1-7)}$$

You can verify this by plotting the slopes at several points on a $\ln x$ versus x graph. This is analogous to taking the derivative.

Since $d/dx(\ln x) = 1/x$ this implies that $\int (1/x)\, dx = \ln x$ which is another special operation which you should remember.

A1-2   Logarithms

DEFINITION OF A LOGARITHM

For each positive number x, the unique number y such that $x = z^y$ is called the logarithm of x to the base z. z must be a positive number other than 0 or 1. We write $y = \log_z x$ (y is the logarithm of x to the base z).

Using this definition of a logarithm it is easy to show that the following combination rules are valid.

1. $\log_z ab = \log_z a + \log_z b$ (A1-8)
2. $\log_z (a/b) = \log_z a - \log_z b$ (A1-9)
3. $\log_z a^n = n \log_z a$ (A1-10)

The most commonly used logarithms are those for which $z = 10$, termed common logarithms, and $z = e$, termed natural logarithms.

LOGARITHMS TO THE BASE e

For each positive number x, the unique number y, in the expression

$$x = e^y$$

is called the "natural or naperian logarithm" of x.

$$\text{Thus } y = \log_e x$$

Note that y is the power to which e must be raised in order to obtain x.

It is customary to write $\log_e x$ (log of x to the base e) as $\ln x$. This helps distinguish natural logarithms from logarithms to the base 10 which are usually denoted by $\log x$. Following are two simple identities which are extremely useful. You will encounter them quite regularly.

1. $e^{\ln x} = x$ (A1-11)

    Let $\ln x = y$

    Then $x = e^y$ from the definition of a logarithm

    Thus $e^{\ln x} = e^y$

               $= x$

2. $\ln e^x = x$ (A1-12)

    $\ln e^x = x \ln e$

    But $\ln e = 1$

    Thus $\ln e^x = x$

If x and a are positive numbers with $a \neq 1$, then

$$x = e^{\ln x} = e^{\ln a \, \frac{\ln x}{\ln a}} = a^{\frac{\ln x}{\ln a}}$$

Therefore,

$$\log_a x = \frac{\ln x}{\ln a} \qquad \text{(A1-13)}$$

If $x = e$, then

$$\log_a e = \frac{\ln e}{\ln a} = \frac{1}{\ln a}$$

$$\log_a x = \log_a e (\ln x)$$

It is often necessary to convert $\ln x$ to $\log x$. Using the Equation (A1-13),

$$\log x = \log e (\ln x)$$

To find $\log e$, turn to tables of Logarithms of Numbers.

$$e = 2.718 \times 10^0$$

So $\log e = (\log 2.718) + 0$. From the tables $\log (2.718) = 0.4343$
Therefore

$$\log x = 0.4343 \, \ln x \qquad \text{(A1-14)}$$

or

$$\ln x = 2.3026 \log x \qquad \text{(A1-15)}$$

A1-3 Graphs

In biophysics we are usually concerned with functional relationships between two (or more) variables. It is important to be able to display these relationships graphically since a graph usually shows the relationship most clearly and vividly. Often in theoretical discussions, relationships are presented in the form of general equations and we should make it a habit to sketch graphs of these equations to "see what they really mean." The following problems will give you some practice in this. It is intended that you obtain the general shape of the graphs and learn how to spot the significance of the constants in the relationships; it is not intended that you draw the graphs accurately by plotting a large number of points. Usually the equations contain constants (symbols other than the variables x and y). If necessary, substitute numbers for these constants until you learn their general significance in each case.

Exercise: On the axes, sketch several graphs of $y = mx + b$ as follows:

(i) m and b both positive numbers
(ii) m as in (i) but b a negative number
(iii) (b) as in (i) but m a negative number
(iv) (b) as in (i) but $m = 0$

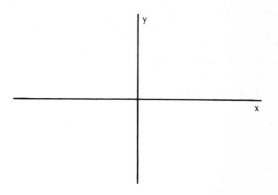

What is the graphical significance of m and b? If m is kept fixed, what is the effect of changing b from $-\infty$ to $+\infty$? If b is kept fixed, what is the effect of varying m from $-\infty$ to $+\infty$?

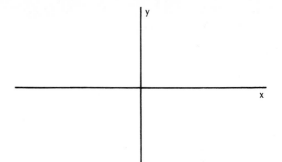

What is special about the case where $b = 0$ i.e. $y = mx$? On the axes sketch this case for

  (i) m is positive and small
 (ii) m is positive and large
(iii) m is negative (any value)

For relationships of this form with m positive, we say that y varies directly or linearly with x.

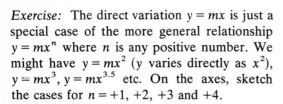

Exercise: The direct variation $y = mx$ is just a special case of the more general relationship $y = mx^n$ where n is any positive number. We might have $y = mx^2$ (y varies directly as x^2), $y = mx^3$, $y = mx^{3.5}$ etc. On the axes, sketch the cases for $n = +1, +2, +3$ and $+4$.

Note the similar shapes for $n > 1$, in the first quadrant (i.e. both x and y positive) these are the cases commonly met in practice.

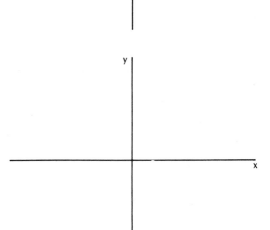

Exercise: Sketch the graph of $y = m/x$ or $xy = m$ where m is a +ve constant. Sketch cases for $m = 1$, $m > 1$ and $m < 1$. What is the significance of changing m? With variations of this type we say that y varies *inversely* as x.

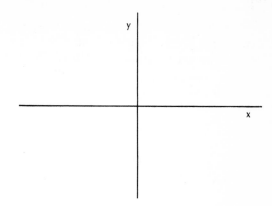

Exercise: The above case, $y = m/x$, is a special case of $y = m/x^n$ where n is a +ve number. i.e. $y = m/x$, $y = m/x^3$ etc.

Sketch curves below for the cases where $n = 1$, 2, 3, and 4 (keep m constant, say $m = 1$).

Note the effect of changing the n and the rather similar shape of all the curves.

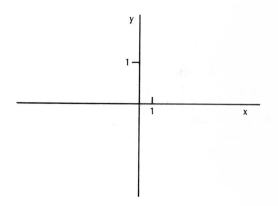

Exercise: An important relationship is the exponential of the general form $y = ae^{\pm kx}$ where a and k are positive constants. To illustrate the significance of the constants sketch the following. (You should be able to obtain approximate curves without using tables if you recall that $e \simeq 3$.)

(i) First sketch the simple cases: $y = e^{+x}$ and $y = e^{-x}$ (i.e. $a = 1$, $k = 1$)

(ii) To see the effect of the constant "a", sketch $y = ae^{+x}$ and $y = ae^{-x}$ for some value of $a \neq 1$. In words, what is the significance of "a"?

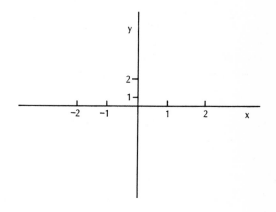

(iii) Finally sketch the general cases $y = ae^{\pm kx}$ for some general value of a and for a very small value of k (say $k = 1/10$) and a very large value of k (say $k = 100$). What is the significance of k?

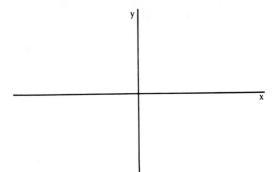

Exercise: Sketch graphs of
(i) $y = \log x$ (i.e. to base 10)
(ii) $y = \ln x$ (i.e. to base e)
Note the similar shape. Do you know the relationship between these two curves?

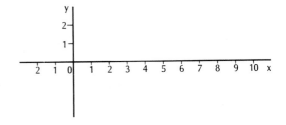

Exercise: Sketch the graph of $y = ae^{-kx^2}$ where a and k are positive constants. This is the famous Gaussian or normal distribution.

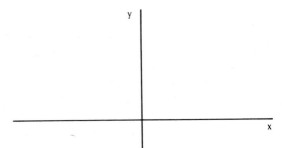

A1-4 Sine and Cosine Functions

In this section we briefly review sine and cosine functions. These functions have an oscillating nature and play an important role in the mathematical description of vibrations and wave motion.

Sine and cosine functions may be defined with reference to Figure A1-1. Consider an x-y coordinate system with a circle of radius r centred at the origin O. A point P moves around the circumference of the circle and the line OP thus sweeps out an angle θ relative to the positive x-axis Ox. (The angle θ is considered positive if measured counter-clockwise from Ox and negative if measured clockwise from Ox.)

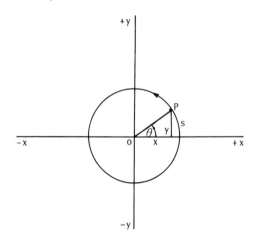

Figure A1-1

Definition of the sine and cosine functions

The angle θ may be measured in degrees or in radians. The measure of an angle in radians is defined by:

$$\theta = s/r$$

where $s =$ the arc length from the x-axis to P and r is the radius OP. An angle is a dimensionless quantity as evident from the above definition of its radian measure.

It is evident from the above definition and the fact that the circumference of the circle $C = 2\pi r$ that $360° = 2\pi$ radians, $180° = \pi$ radians, $90° = \pi/2$ radians and 1 radian $= 57.3°$; thus it is easy to convert from radians to degrees by these relationships.

The sine of the angle θ (written $\sin \theta$) is the dimensionless ratio y/r

$$\text{i.e. } \sin \theta = y/r$$

where y is the ordinate of P.

Similarly the cosine of the angle θ (written $\cos \theta$) is the ratio x/r

$$\text{i.e. } \cos \theta = x/r$$

where x is the abscissa of P.

When applying these definitions, the sign of x and y must be considered; r is positive for all angles.

The way in which $\sin \theta$ varies with θ is shown in the graph of Figure A1-2a where θ is plotted in both degrees and radians over a range of three complete revolutions of OP, i.e. from $\theta = -360°$ to $+720°$. The sine is 0 at those angles where y is zero, i.e. $\theta = 0°$, $\pm180°$, $\pm360°$ etc. Sin θ is positive where y is positive, negative when y is negative, i.e. when P lies below the x-axis. The extreme values of $\sin \theta$ are ±1, occurring at $\theta = 90°$, $270°$ etc. Where P crosses the y-axis and $y = +r$ or $-r$.

The variation of $\cos \theta$ with θ is given in Figure A1-2b. The cosine function also oscillates between ±1 and is identical to the sine function except for a shift of $\pi/2$ radians along the θ-axis.

The numerical values of $\sin \theta$ and $\cos \theta$ are given in standard tables for the range of θ from $0°$ to $+90°$. Values for any angle outside this range can easily be related to the tabulated values from the symmetry of the above graphs (see examples below). For negative angles it is useful to note from the graphs for

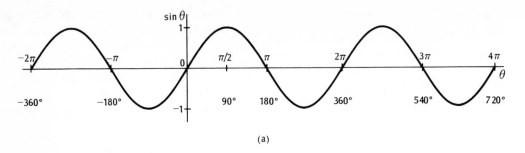

(a)

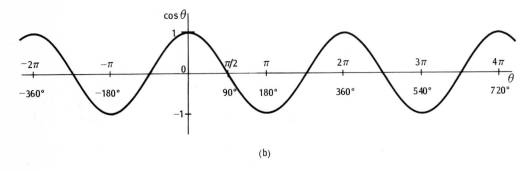

(b)

Figure A1-2

Graphs of $\sin \theta$ and $\cos \theta$

any angle θ, $\sin (-\theta) = -\sin (+\theta)$ and $\cos (-\theta) = +\cos (+\theta)$.

EXAMPLE

Evaluate $\sin 300°$ using a standard sine table. From the symmetry of the $\sin \theta$ vs. θ graph it is evident that $\sin 300° = \sin (360 - 60°) = -\sin 60°$. $\sin 60° = 0.866$ from the standard sine table

$$\therefore \sin 300° = -0.866$$

EXAMPLE

Evaluate $\sin (-4.00)$ where -4.00 is an angle in radians. First, convert the function to a function of a positive angle. Since $\sin (-\theta) = -\sin \theta \therefore \sin (-4.00) = -\sin 4.00$. Convert

the radian measure to degrees. Since 2π radians $= 360°$

$$\therefore 4.00 \text{ radians} = \frac{360°}{2\pi} \times 4.00 = 229°$$

From the symmetry of the sine function, $\sin 229° = \sin (180° + 49°) = -\sin 49°$ $\therefore \sin 229° = -\sin 49° = -0.755$ from the standard sine table.

and $\therefore \sin (-4.00) = -\sin 229°$
$$= -(-0.755)$$
$$= +0.755$$

EXAMPLE

Evaluate $\cos 920°$. The sine and cosine functions repeat each $360°$. Therefore, subtract the largest possible integral multiple of $360°$ to find the angle between 0 and $360°$ which has the same cosine as $920°$. In this case:

$\cos 920° = \cos (920°-2\times360°) = \cos 200°$.

From the $\cos \theta$ vs. θ graph

$$\cos 200° = \cos (180°+20°)$$
$$= -\cos 20°$$
$$= 0.940$$
$$\therefore \cos 920° = 0.940$$

Some useful trigonometric identities which may be readily verified are:

(i) $\sin^2 \theta + \cos^2 \theta = 1$ (A1-16)

(ii) $\sin (\alpha \pm \beta) = \sin \alpha \cos \beta \pm \cos \alpha \sin \beta$

(A1-17)

(iii) $\cos (\alpha \pm \beta) = \cos \alpha \cos \beta \mp \sin \alpha \sin \beta$

(A1-18)

The symbol $\sin^2 \theta$ in (i) above means $(\sin \theta)^2$. For example, $\sin^2 300° = (\sin 300°)^2 = (-0.866)^2 = +0.750$

Exercise: Evaluate the following using standard sine and cosine tables:

(i) $\sin 76.2°$ (ii) $\cos (1.3r)$ (iii) $\cos (-170°)$ (iv) $\sin (1000°)$ (v) $\sin (10r)$.

(The symbol "r" indicates radian measure.)

Exercise: Sinusoidal relations of the form $y = y_o \sin kx$ or $y = y_o \cos kx$ are important in all vibration and wave phenomena and in wave mechanics.

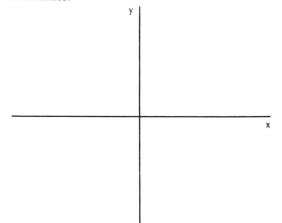

(i) first sketch the simple cases $y = \sin x$ and $y = \cos x$ (i.e. $y_o = 1$ and $k = 1$). Think of x as an angle in radians.

(ii) On the same axes sketch $y = y_o \sin x$ for $y_o \neq 1$. What is the significance of y_o?

(iii) Again sketch $y = y_o \sin x$ on these new axes. Now on the same axes sketch $y = y_o \sin kx$ for $k > 1$ (say $k = 2$). What is the significance of k?

Problems

1. (a) $\log 5.7 =$ (d) $\log (1/5.7) =$
   (b) $\log 57 \ =$ (e) $\log (5.7)^{5.7} =$
   (c) $\log 0.57 =$ (f) $\ln 5.7 \ \ \ =$

2. Find $\ln x$ when:
   (a) $x = (27)^{-2}$
   (b) $x = 4.2 \times 10^{-3}$

3. Find $\log x$ when:
   (a) $x = e^2 \times 0.026$
   (b) $x = 0.362$

4. Find x when:
   (a) $\log x = \bar{1}.7$
   (b) $\log x = -0.3$
   (c) $x = e^4 \times e$
   (d) $x = (35)^{1/2}/2.15$

5. If $\log_{10}x + \log_{10}x^2 + \log_{10}x^3 + \log_{10}x^4 = 10$, what is the value of x?

6. Prove that $\log_{100} N = \frac{1}{2} \log N$.

7. Assume that the earth's population is 7 billion people and that from past statistics, the doubling time for this population is 35 years (assuming exponential growth). What is the growth rate constant for this population?

8. An exponentially decreasing population is halved after 10 minutes. How much longer will it be until the population is one third its original size?

9. The number of cells in a bacterial culture at the time of preparation was 10^3. At

6:00 p.m. the population numbered 10^6 cells and one hour later it had doubled to 2×10^6 cells. What was the time of preparation?

10. A cell culture has a growth constant $k = 0.5\,\mathrm{hr}^{-1}$. By what factor will the initial size of the culture increase in two hours?

11. Three cultures of different cells start off each with the same number of cells and decay with different rate constants. The magnitude of the rate constants are:

$$\begin{aligned}
&\text{1st culture}\quad k = 0.005\,\mathrm{s}^{-1}\\
&\text{2nd culture}\quad k = 0.008\,\mathrm{s}^{-1}\\
&\text{3rd culture}\quad k = 0.010\,\mathrm{s}^{-1}
\end{aligned}$$

Which culture will decrease to 10 percent of its initial size first?

12. A wealthy man wants to have a party at his Northern Ontario lodge in May, which is the worst month of the year for bugs. He figures there are 10,000 times too many black-flies and 100 times too many mosquitoes to be comfortable. A certain pesticide will kill black-flies so they die off with a decay constant of $-3.1\,\mathrm{days}^{-1}$ and it will kill mosquitoes with a decay constant of $-1.15\,\mathrm{days}^{-1}$. If the party is to be on May 10, what is the latest date that the pesticide could be sprayed so that both insects will be at a comfortable level for the party?

13. A phage (bacterial virus) attacks a bacterial cell by injecting its own genetic material into the bacteria and using the cell's

biochemical machinery during the creation of more virus particles. If a bacterium bursts, spilling 100 new virus particles 25 minutes after infection by a single virus, what is the exponential growth rate constant for this virus in a bacterial colony if all the released virus particles are successful in infecting other bacteria?

14. Given the equation $s = x - \frac{1}{2}at^2$, where x and a are positive constants, sketch a graph (for $-\infty < t < +\infty$) of
 (a) s vs. t
 (b) s vs. t^2

15. For the equation $V = \frac{4}{3}\pi r^3$, sketch a graph (for $r \ge 0$) of
 (a) V vs. r
 (b) V vs. the power of "r" that gives a straight line.

16. What is the significance of "a" in the function $y = ae^{-kx}$?

17. Sketch the graphs of the following equations. Show the coordinate axes with some indication of scale. Give the coordinates of at least 2 points on the graph.
 (a) $y = 2\sin\left(\dfrac{\pi}{2} - x\right)$
 (b) $y = e^{(x/500)}$
 (c) $y = x^3$
 (d) $y = \dfrac{2}{x}$

18. Prove, using a diagram, that for any angle $\sin^2\theta + \cos^2\theta = 1$.

19. Without using tables, evaluate $\sin 60°$, $\cos 30°$, $\cos 45°$, $\sin 45°$.

20. Show graphically that $\sin(90° + \theta) = \cos\theta$.

Appendix 2

Units and
Dimensional Analysis

A2-1 Introduction

In this section we present a rather heterogeneous collection of facts concerning the units and dimensions of physical quantities.

Most physical quantities can be expressed in terms of a few fundamental quantities such as length (L), time (T), mass (M), current (I), and temperature (θ). For example, by its definition, velocity is a "length" divided by a "time interval" i.e. velocity = length/time or L/T or LT^{-1}. We say that velocity has *dimensions* of "length $\times$ time^{-1}" which we write simply as LT^{-1}. Similarly acceleration has dimensions of velocity/time or LT^{-1}/T or LT^{-2}.

Any quantity with dimensions also requires *units* to express its magnitude. Many different units may be used to measure the same quantity. For example velocity is commonly expressed in miles hr^{-1}, ft s^{-1}, cm s^{-1}, m s^{-1}, etc; however velocity has only one dimension, namely LT^{-1}. (See Appendix 4.)

Note that the usual rules for working with exponents apply to dimensions (eg. $LT^{-1}/T = LT^{-2}$ (as above)) and units (e.g. m s$^{-2} \times$ s$^2 =$ m).

Some of the quantities we use are dimensionless. For example "pure" numbers like the trigonometric functions, the logarithm of any number and exponential "e" raised to any power (i.e. e^x) are dimensionless. Quantities which we simply count are dimensionless. Thus, although we express a vibrational frequency in "cycles per second" the number of cycles is dimensionless but the per second gives to the frequency the dimensions of T^{-1}.

Quantities which have dimensions may yield dimensionless quantities if properly combined. For example the product "frequency $\times$ time" is dimensionless i.e. $T^{-1} \times T = T^o = 1$, a pure number. Ratios of "like" quantities such as "length/length" are dimensionless. For example π (circumference/diameter), sines, cosines etc. and the radian measure of an angle (arc/radius) are all dimensionless.

A2-2 Conversion of Units

One of the unfortunate facts of life in biophysics is the overabundance of units used to express the magnitude of the same physical quantity. In this book, we have tried to consistently apply the S.I. rules for units. This system is very similar to the M.K.S. (metre, kilogram, second) system which preceded it. Many extremely good books use still other systems. It is, therefore, extremely important for all people in science to be able to convert rapidly from one set of units to another. It is easy enough to look up simple conversion factors in tables but we may easily make mistakes in complicated conversions because we divide when we should multiply or vice versa. The following examples illustrate a foolproof method relying on the cancellation of units. The method involves multiplying the conversion factors which are written as ratios in a way that the unwanted units cancel each other and the desired units remain. As long as the ratios are written in this manner the correct conversion will be made (excluding arithmetic errors).

EXAMPLE

Convert 2.0 calories to joules. Conversion factor tables give: 1 cal = 4.186 J

$$\therefore 2.0 \, cal = 2.0 \, cal \times \frac{4.186 \, J}{1 \, cal}$$

$$= (2.0 \times 4.186) \, J$$

$$= 8.4 \, J$$

Convert 3.0 joules to calories

$$3.0 \, J = 3.0 \, J \times \frac{1 \, cal}{4.186 \, J}$$

$$= \frac{3.0 \, cal}{4.186}$$

$$= 0.72 \, cal$$

EXAMPLE

Convert 2.0 calories to electron-volts (eV). Conversion factor tables give 1 cal = 4.186 J

$$1 \, J = 10^7 \, erg$$

$$1 \, eV = 1.60 \times 10^{-12} \, erg$$

$$\therefore 2.0 \, cal = 2.0 \, cal \times \frac{4.186 \, J}{1 \, cal}$$

$$\times \frac{10^7 \, erg}{1 \, J} \times \frac{1 \, eV}{1.60 \times 10^{-12} \, erg}$$

$$= \left(\frac{2.0 \times 4.186 \times 10^7}{1.60 \times 10^{-12}}\right) eV$$

$$= 5.2 \times 10^{19} \, eV$$

EXAMPLE

The method may be extended to quantities involving mixtures of units as illustrated

below. The dissociation of H_2 into H atoms requires 103 kcal mol^{-1}. What is this in eV molecule^{-1}? Use the conversion factors cited above, plus the fact that 1 kcal = 10^3 cal and Avogadros number = 6.0×10^{23} molecules/mole.

$$103 \frac{kcal}{mole} = 103 \frac{kcal}{mole} \times \frac{10^3 \, cal}{1 \, kcal}$$

$$\times \frac{4.186 \, J}{1 \, cal} \times \frac{10^7 \, erg}{1 \, J}$$

$$\times \frac{1 \, eV}{1.60 \times 10^{-12} \, erg}$$

$$\times \frac{1 \, mole}{6.02 \times 10^{23} \, molecule}$$

$$= \left(\frac{103 \times 4.186}{1.60 \times 6.02} \times 10^{-1}\right) eV \, molecule^{-1}$$

$$= 4.48 \, eV \, molecule^{-1}$$

EXAMPLE

Express the area 2.0 ft^2 in units of cm^2 using the facts that 1 inch = 2.54 cm and 1 foot = 12 inches.

$$2.0 \, ft^2 = 2.0 \, ft^2 \times \left(\frac{12 \, in}{1 \, ft}\right)^2 \times \left(\frac{2.54 \, cm}{1 \, in}\right)^2$$

$$= (2.0 \times 12^2 \times 2.54^2) \, cm$$

$$= 1860 \, cm^2$$

Exercise: Verify the equalities listed below. Use your knowledge of the metric system, common units of length and time and any of the conversion factors listed in the above examples.

(i) 60 mi hr^{-1} = 88 ft s^{-1}
(ii) 1 ft^3 = 28.3 litres
(iii) 1.0 eV molecule^{-1} = 23 kcal mol^{-1}
(iv) 60 ergs cm^{-2} = 38 × 10^{-4} ev A^{-2} (1A = 1 Angstrom unit = 10^{-8} cm)

A2-3 Dimensional Equivalence

A given physical quantity is sometimes expressed in apparently different sets of units. For example, the surface tension of a liquid is expressed as "force per unit length" (e.g. $N\,m^{-1}$) and as "energy per unit area" (e.g. $J\,m^{-2}$). Dimensional analysis is useful in showing the equivalence of such expressions. One simply shows that the alternate expressions have the same dimensions. In the case of surface tension:

$$\frac{force}{length} = \frac{mass \times acceleration}{length}$$

$$= \frac{M \times LT^{-2}}{L} = \frac{M}{T^2}$$

$$\frac{energy}{area} = \frac{force \times length}{area}$$

$$= \frac{M \times LT^{-2} \times L}{L^2} = \frac{M}{T^2}$$

which shows the equivalence.

Often it is not necessary to work out the dimensions in terms of the fundamental quantities (mass, etc.).

In the above case:

$$\frac{energy}{area} = \frac{force \times length}{(length)^2} = \frac{force}{length}$$

which shows the equivalence directly.

Exercise: (i) Show that "pressure × volume" is dimensionally equivalent to work or energy. This equivalence is often apparent in thermodynamic expressions.

(ii) The electric field intensity (E) at a point is defined as the force per unit charge placed at the point (i.e. $E = F/q$). However, one often finds electric fields expressed in "volts/metre" (called the electrical potential gradient at the point). Show that "newtons/coulomb" is dimensionally equivalent to "volts/metre".

Hint: recall that a volt is a "joule/coulomb".

(iii) Bernoulli's equation states that at different points in a flowing liquid, the sum "$P + \rho g y + \frac{1}{2}\rho v^2$" is constant. Here P = pressure in the liquid at the point, ρ = density of the liquid, g = acceleration due to gravity, y = height of the point in question above a reference level, and v = velocity of the liquid at the point. Show that each term p, $\rho g y$ and $\frac{1}{2}\rho v^2$ has the same dimensions. They must be equivalent if they can be added together.

Exercise: In any equation, each term must have the same dimensions (and the same units when physical quantities are substituted into the equation). Checking for dimensional consistency can often show up errors in an equation. Check the following equations to see if the dimensions of each term on the left of the equation are equal to the dimensions of each term on the right. (At least one equation is incorrect. Can you correct it (them)?). Note in case (i) $v_0 t$, $\frac{1}{2}at^2$ are each "terms", in (ii) "$y_o \sin(\omega t - kx)$" is all one term.

(i) The equation for rectilinear motion with constant acceleration: $x = x_0 + v_0 t + \frac{1}{2}at^2$

where x = position at time t
$\quad a$ = acceleration
$\quad v_0$ = velocity at $t = 0$
$\quad x_0$ = position at $t = 0$

(ii) The equation for a one-dimensional travelling wave:

$$y = y_o \sin(\omega t - kx)$$

where y = displacement from equilibrium of a particle at time t
$\quad y_0$ = maximum displacement of the particle
$\quad x$ = equilibrium position of particle
$\quad \omega = 2\pi \nu$ where ν is the frequency of the wave

and $\quad k = 2\pi/\lambda$ where λ is the wavelength

(iii) The expression for the frequency of a simple pendulum

$$v = \frac{1}{2\pi}\sqrt{\frac{L}{g}}$$

where v = frequency
L = length of pendulum
g = acceleration due to gravity

(iv) The growth equation

$$ln\frac{N}{N_0} = kt$$

Exercise: Proportionality constants are often encountered in physical equations. These constants must assume dimensions and units such that the equation will be dimensionally correct and have the same units on each side of the equality.

e.g. Newton's Law of Gravitation states that the gravitational force (F) between two point masses m_1 and m_2 varies directly as their product and inversely as the square of their separation r. i.e.

$$F \propto \frac{m_1 m_2}{r^2}$$

or $F = G(m_1 m_2 / r^2)$ where G is a proportionality constant. What are the dimensions of G and its units? The simplest way to answer this is to simply solve for G as an algebraic quantity

$$G = \frac{Fr^2}{m_1 m_2}$$

Therefore its dimensions are:

$$\frac{\text{Force} \times \text{length}^2}{\text{mass}^2} = \frac{M \times L \times L^2}{T^2 \times M^2} = \frac{L^3}{MT^2}$$

and its units are

$$\frac{\text{N m}^2}{\text{kg}^2} = \frac{\text{m}^3}{\text{kg s}^2}$$

(i) What are the dimensions and units of the proportionality constant k in Coulomb's law?

$$F = kqq'/r^2$$

where F is the force between charges q and q' a distance r apart? The coulomb is the unit of charge.

(ii) What are the dimensions and units of the gas constant R in the ideal gas law

$$PV = nRT$$
P = pressure
V = volume
n = number of moles
T = absolute temperature

(iii) A table of fundamental constants gives "R" in litre atmospheres $\text{mol}^{-1}\text{K}^{-1}$. Are these possible units for R? If so, what units must be used for P, V, n and T in the gas law?

A2-4 Dimensions and Transcendental Functions (sines, logs, exponentials)

An expression such as "log Z" has meaning only if Z is a dimensionless quantity. For example, there is meaning to log (32) but not log (32 ft s^{-2}). Similarly sin Z and e^z have meaning only if Z is dimensionless (in the case of sin Z, cos Z, etc., if Z is any dimensionless quantity, we can always think of Z as an angle in radians). In these expressions, the quantity Z is called the argument of the function. In physical equations involving these functions, the arguments usually involve physical quantities which have dimensions; if so, these must be so arranged as to leave the argument as a whole, dimensionless. For example, the equation giving the displacement from equilibrium x, of a particle oscillating in simple harmonic motion may have the form: $x = x_o \sin(2\pi vt)$ where x_o is the maximum displacement, v is the frequency and t is the time.

The product "$2\pi vt$" is the argument of the sine and corresponds to the Z of the above discussion. If this expression is correct, $(2\pi vt)$ should be dimensionless; examination of the individual factors shows that it is $T^{-1} \times T = 1$.

Exercise: (i) The general expression for a travelling one-dimensional sinusoidal wave may be written in the form:

$$y = y_o \sin(\omega t - kx)$$

where y is the displacement from equilibrium at time t of a particle whose equilibrium position is x. In this expression, ω and k are constants. What must be the dimensions of ω and k so that the argument of the sine will be dimensionless (each term of the argument must be dimensionless)?

(ii) The expression for the decay of a radioactive element has the form $N = N_o e^{-\lambda t}$ where:

N is the activity of the material at time t.
N_o is the activity at $t = 0$ and λ is a constant depending on the material. (known as the "decay constant"). What are the dimensions of λ? What are its units?

(iii) Beer's law states that the intensity of light (I) after passing through a thickness x of a solution is given by:

$$I = I_o e^{-\varepsilon cx}$$

where I_o = intensity at $x = 0$, c = concentration of solution, ε = a constant called the extinction coefficient which depends on the solute and the wavelength of the light. What are the units of ε if x is in cm and c is in mol l^{-1}?

Exercise: Complete the following chart giving the dimensions and units requested.

Quantity	Dimensions	SI Units
Force (recall $F = ma$)		
Energy or work ($W = Fs \cos \theta$)		
Pressure ($P = F/A$)		
Electric field ($E = F/q$)		
frequency of a sound wave		
number of molecules per unit volume		
log 5.6		
sin 30		
$e^{2.4}$		
radian measure of 30°		

Problems

1. The physical quantity viscosity is defined as a force per (unit area per second). Derive the dimensions of viscosity and give its units.
2. Some biophysics students who were interested in blood flow decided to try their hand at theoretical physics and they "derived" the following equations relating the velocity of flow (v) to the blood vessel radius (R), the blood pressure (P), the blood density (ρ) and the acceleration due to gravity (g). Unfortunately only one of their equations is dimensionally correct. Which one is it?

(a) $v = \frac{1}{2}g \sin RP$

(b) $v = \sqrt{\dfrac{P}{\rho}}$

(c) $v = \rho gR$
(d) $v = g \ln R/P$
(e) $v = P\rho$

3. The Boltzmann distribution law is:

$$N_r = N_o e^{-E_r/kT}$$

where E_r is an energy term and T is temperature (K). What would be the units of k?

4. Which of the following is *not* an energy unit?

(a) newton (b) joule (c) erg (d) electron-volt (e) calorie

5. The wavelength of an electron λ_e is given by the following equation

$$\lambda_e = \frac{h}{m_e v}$$

v is the velocity of the electron. What are the dimensions of h (Planck's constant), and its units?

Appendix 3
Vibrations
and Waves

A3-1   Introduction

Vibrations and waves are among the most important types of motion found in nature. The movement of an insect's wing during flight and the oscillations of the tympanic membrane of the ear are examples of vibrational motions of interest in biology. Wave phenomena occur in studies of sound, light, and the energy and arrangement of electrons in atoms and molecules. In this appendix, we present some of the terms and mathematics used to describe vibrational and wave motion. Many applications to living systems are presented in the text.

A3-2   Simple Harmonic Motion

Many objects in nature, ranging from atoms in molecules to children's swings, are subject, when displaced slightly from their equilibrium position, to a linear restoring force. Such objects, if displaced slightly from equilibrium and released, will undergo a sinusoidal, free oscillation known as simple harmonic motion. A simple, although somewhat academic model of such a system is shown in Figure A3-1. A mass m rests on a horizontal frictionless surface. The mass m is attached to a spring S whose mass will be assumed to be negligible. When the mass is at position 0, the equilibrium position, the spring exerts no force on it. If the mass is displaced to the right ($+x$ direc-

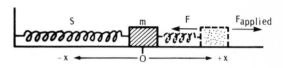

Figure A3-1
Mass spring system executing simple harmonic motion

tion) or to the left ($-x$ direction) the spring is stretched or compressed and exerts a force F on the mass in the direction opposite to x, i.e. back towards 0 (a "restoring" force). For small displacements, it is found that the magnitude of F is proportional to the displacement x; such a force is called a *linear* restoring force. This behaviour is summarized in the graph of Figure A3-2 and in Equation (A3-1).

$$F = -kx \qquad \text{(A3-1)}$$

The negative sign of Equation (A3-1) indicates that F and x are in opposite directions. The proportionality constant "k" of Equation (A3-1) is known as the "force constant" of the system and represents the force per unit elongation or compression of the spring. From Equation (A3-1) it is evident that "k" has dimensions of MT^{-2} and units of $N\,m^{-1}$. A stiff spring has a large force constant, a soft spring has a small force constant.

The system of Figure A3-1 may be described in terms of its potential energy. To

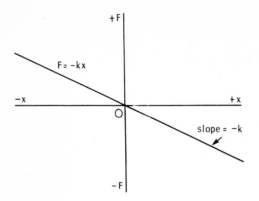

Figure A3-2
The restoring force as a function of displacement for simple harmonic motion

displace the mass m at constant speed you must do work against the spring. You must exert an applied force F exactly equal in magnitude but opposite in direction to that of the spring as in Figure A3-3. (F applied $= +kx$). The work you do in stretching the spring from 0 to $+x$, or in compressing it from 0 to $-x$, is equal to the area under the F applied vs. x curve (the shaded triangle) of Figure A3-3. From Figure A3-3 it is evident that this work is equal to $(\frac{1}{2}x)(kx) = (\frac{1}{2}kx^2)$ and is positive for either elongation or compression of the spring. This work may be considered as potential energy E_P stored in the spring and may be converted to other forms of energy such as kinetic energy when the spring returns to its original length. Thus for a linear restor-

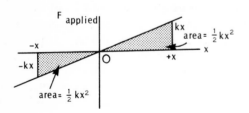

Figure A3-3
Evaluation of the work done on the spring

ing force the potential energy of the system is given by:

$$E_P = \tfrac{1}{2}kx^2 \qquad (A3\text{-}2)$$

The graph of E_P versus x (Figure A3-4) is a parabola and the mass m is said to oscillate in a "parabolic potential well."

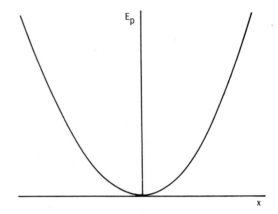

Figure A3-4
Parabolic potential well for a simple harmonic oscillator

If the mass of Figure A3-1 is pulled aside from 0 to x_o and released, how will its position x vary with time t? In a truly frictionless system, the mass will oscillate indefinitely about 0 between the positions $+x_o$ and $-x_o$. Application of Newton's second law of motion tells us that the particle will oscillate sinusoidally with time. If we choose to let $t = 0$ at the instant when the particle passes through the equilibrium point ($x = 0$) travelling in the $+x$ direction, then a graph of its position x versus time t will be a sine curve as in Figure A3-5. The equation of this curve is

$$x = x_o \sin\left(\frac{2\pi t}{T}\right) \qquad (A3\text{-}3)$$

and can be used to calculate exactly where the particle is (x) at any time (t).

The motion is characterized by two constant

quantities, the *amplitude* x_o and the *period T*. The amplitude x_o is the absolute value of the two extreme positions of the particle, that is the particle oscillates between $+x_o$ and $-x_o$. The period T (usually expressed in seconds) is the time for one *complete* oscillation or cycle of the particle; for example the time to travel from $+x_o$ to $-x_o$ and back to $+x_o$, as shown in Figure A3-5.

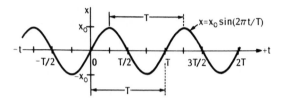

Figure A3-5
Displacement of a simple harmonic oscillator as a function of time

The number of complete oscillations per second is called the *frequency* ν and is obviously given by the reciprocal of the period T.

$$\nu = \frac{1}{T} \qquad \text{(A3-4)}$$

For example, if the period of an oscillation is 1/5 second, the particle will make five complete oscillations per second, and the frequency ν is $5\ \text{s}^{-1}$. Note that the units of frequency are s^{-1}. This unit is given the name "hertz" in honour of one of the original workers in the field of electromagnetic waves. Thus we might say that the frequency in the above case is 5 hertz.

As one might expect, the period and frequency of an oscillator depend on its mass m and the force constant, k of the system. It can be shown that

$$\nu = \frac{1}{2\pi}\sqrt{\frac{k}{m}} \quad \text{and} \quad T = 2\pi\sqrt{\frac{m}{k}} \qquad \text{(A3-5)}$$

The student should check that Equation (A3-5) is dimensionally correct.

From the above discussion it is evident that we could write Equation (A3-3) in terms of the frequency ν, i.e.

$$x = x_o \sin\left(2\pi\nu t\right) \qquad \text{(A3-6)}$$

The quantity "$2\pi\nu$" in Equation (A3-6) is often called the "angular frequency" of the motion and is designated by the symbol "ω". The units of ω are s^{-1} since it is simply frequency multiplied by the dimensionless constant 2π. Obviously from Equation (A3-5), $\omega = \sqrt{k/m}$.

Thus the relation between x and t may also be written:

$$x = x_o \sin\left(\omega t\right) \qquad \text{(A3-7)}$$

The sine function is defined for an angle θ and an angle is a dimensionless quantity. The various forms of Equation (A3-3) do not violate this concept since the quantities $[(2\pi/T)t]$, $(2\pi\nu t)$ or (ωt) are dimensionless ratios and are angles expressed in radians. The quantities $2\pi/T$, $2\pi\nu$ or ω may be considered as factors which convert the variable time to a variable angle which increases linearly with time t.

Of course, one can choose to let $t = 0$ at any point in the cyclic motion of the particle. The general graph of x vs. t will still be sinusoidal as in Figure A3-5 but with the sine curve shifted anywhere up to $\frac{1}{2}$ period $(T/2)$ to left or right of the time origin, the exact distance depending on just where in the cycle you choose to let $t = 0$. The equation of such a shifted sine curve is

$$x = x_o \sin\left(\omega t + \delta\right) \qquad \text{(A3-8)}$$

where δ is a dimensionless constant called the phase angle, the exact value of which depends on the point in the cycle where $t = 0$, i.e. it depends on the amount of shift in the sine curve. It is normally simplest to use the $t = 0$ conditions of Figure A3-5 in which case $\delta = 0$.

Another simple choice is to let $t = 0$ when the particle is at $x = x_o$. The graph of x vs. t is then a standard cosine curve (i.e. the sine curve of Figure A3-5 shifted $\frac{1}{4}T$ to the left). The analytical relation between x and t is then $x = x_o \cos \omega t$ which can also be written $x = x_o \sin\left(\omega t + \frac{\pi}{2}\right)$. Particles which oscillate with the same period and pass through zero together, reach their positive maximum displacement together (as particles 1 and 2 of Figure A3-6a) are said to oscillate "in phase" since their phase angles are always identical. However if one particle is at its positive maximum displacement when the other is at its negative maximum, as in Figure A3-6b their phase angles will differ by π radians or 180 degrees and they are said to oscillate 180 degrees or π radians out of phase. All phase differences from 0 to $\pm\pi$ radians are possible.

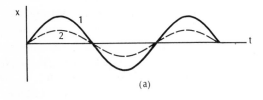

(a)

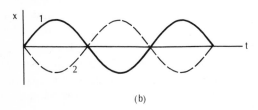

(b)

Figure A3-6
(a) Particles oscillating in phase
(b) Particles oscillating 180° out of phase

As a particle oscillates back and forth, its velocity changes with time and with position x. It is easy to show by differentiating Equation (A3-7) with respect to time (since $v = dx/dt$)

that the velocity varies sinusoidally with time, being (of course) zero at those times when $x = \pm x_o$ and a maximum $(\pm 2\pi x_o/T)$ when the particle passes through $x = 0$ (you travel fastest on a swing as you pass through the central part). Similarly the acceleration (found from $a = dv/dt$) also varies sinusoidally with time and is zero as the particle passes through $x = 0$ and a maximum $(\mp 4\pi^2 x_o/T)$ at the extreme positions $x = \pm x_o$. This is, of course, in agreement with Newton's laws of motion and the linear restoring force relationship $F = -kx$ (acceleration will be zero when $F = 0$ (or $x = 0$) and a maximum where F is a maximum).

EXAMPLE

(a) The spring in Figure A3-1 exerts a restoring force of 5 N when stretched or compressed 1/5 metre. If a 1 kg mass is attached to the spring, at what frequency and period will the system oscillate?

The force constant

$$k = -F/x = -(-5 \text{ N})/(+1/5 \text{ m}) = 25 \text{ N m}^{-1}$$

Then for a mass of 1 kg

$$\nu = \frac{1}{2\pi} \sqrt{\frac{k}{m}} = \frac{1}{2\pi} \sqrt{\frac{25 \text{ N m}^{-1}}{1 \text{ kg}}}$$

$$= \frac{5}{2\pi} \text{s}^{-1} = 0.8 \text{ s}^{-1}$$

$$T = \frac{1}{\nu} = \frac{2\pi}{5} \text{s} = 1.26 \text{ s}$$

The angular frequency $\omega = 2\pi\nu = 5 \text{ s}^{-1}$
(b) If the above mass is pulled aside 0.02 metres and released, what is the equation relating x and t if you let $t = 0$ as the particle passes through $x = 0$ travelling in the $+x$ direction?

For these conditions: $x = x_o \sin \omega t$

$$x = 0.02 \sin 5t$$

where x is in metres and t is in seconds.

(c) Sketch the motion – i.e. sketch a graph of x vs. t for the $t = 0$ conditions of (b) above.

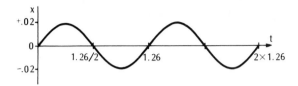

(d) What is the displacement x at the instant $t = \frac{1}{2}$ s?

Since $x = 0.02 \sin 5t$

$$x = 0.02 \sin (5 \times \tfrac{1}{2}) \text{ at } t = \tfrac{1}{2} \text{ s}$$

$$x = 0.02 \sin (2.5 \text{ radians})$$

$$x = 0.02 \sin (143°)$$

$$= 0.02 \sin (180° - 37°)$$

$$= 0.02 \sin (37°)$$

$$= 0.02 \times 0.602$$

$$= 0.012 \text{ m}$$

EXAMPLE

The equation $x = 2 \sin (3t + 0.4)$ describes the simple harmonic motion of a particle where x is in metres and t is in seconds. What is the amplitude, frequency, and period of the motion?

By comparison with the general equation for simple harmonic motion $x = x_o \sin (\omega t + \delta)$ it is obvious that the amplitude is 2 m and the angular frequency is 3 s^{-1}.

$$\omega = 2\pi\nu = 3 \text{ s}^{-1}$$

$$\nu = \frac{3}{2\pi} \text{ s}^{-1} = 0.48 \text{ s}^{-1}$$

and

$$T = \frac{1}{\nu} = \frac{2\pi}{3} \text{ s} = 2.1 \text{ s}$$

In any real macroscopic system, there is of course some friction. Thus such systems do not oscillate indefinitely if simply pulled aside and released. The friction causes the amplitude to decrease with time although it has very little effect on the period. This is called "damped" harmonic motion and is illustrated in Figure A3-7a. If the friction is very large relative to the restoring force, the particle will not oscillate at all but simply return slowly to its equilibrium point as in Figure A3-7b. It is only in atomic systems that friction is absent and we can observe simple harmonic motions which continue indefinitely.

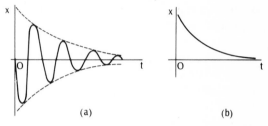

Figure A3-7
Damped harmonic motion

The oscillations we have been discussing are free oscillations of a system at the frequency determined by its mechanical properties, as given by Equation (A3-5). Of course it is possible to attempt to drive a system into oscillation at any frequency by applying a driving force to it which varies sinusoidally with time. When one tries this, one finds that it is quite difficult to force the system to vibrate with large amplitude except at or very near to its free oscillation frequency i.e. the frequency given by Equation (A3-5). However if you do drive it at its free oscillation frequency, oscillations of very large amplitude can easily be built up. This phenomenon is called resonance and the free oscillation frequency is often called the resonance frequency. A child's swing is a common example. It has a natural oscillating or swinging frequency and it is hard to push it back and forth at any other frequency. However, gentle pushes applied at the natural oscillating frequency can very easily cause the swing to swing back and forth with a very large amplitude. These resonance phenomena are very important in nature.

A-3 Travelling Waves

We are all familiar with waves travelling across the surface of the water. In such waves the disturbance (i.e. energy) moves forward across the water although the water particles (except where the water is very shallow) move approximately up and down at one location. Thus the water does not move forward although the wave or disturbance does. This may be verified by watching a floating object bob up and down at one location as waves move forward beneath it. A travelling wave may be defined as a disturbance which moves through space carrying energy without the bulk forward movement of matter. The water particles move of course; they oscillate about a mean position but do not move forward with the wave.

Sound is another example of wave motion and is discussed in more detail in Chapter 2. Sound waves differ from water waves in one important way. In water waves, the particles oscillate in a direction perpendicular (vertical) to the direction that the wave travels (horizontal). Such waves are called transverse waves. On the other hand, in a sound wave, the particles oscillate back and forth in the same direction as the wave as a whole travels. Such waves are called longitudinal waves. Light consists of transverse oscillations of electric and magnetic fields and is therefore called an electromagnetic wave. These waves are discussed in detail in Chapter 3. They differ from sound and water waves in that the oscillations of material particles are not involved but rather the oscillation of fields (electric and magnetic). As such they can readily carry energy through space completely void of material particles.

Because transverse waves have associated with them some unique direction (such as the plane of oscillation) perpendicular to their direction of travel, they exhibit a whole set of phenomena known as polarization phenomena not found with longitudinal waves. For example light, which is a transverse wave, exhibits well-known polarization effects. Light will not pass through a sheet of polaroid (as in polaroid sunglasses) if the electric field oscillations in the light wave are parallel to the direction of the many parallel, long, chain-like organic molecules in the polaroid. If so, the oscillating electric field causes some of the electrons in the molecules to move along the chains and the wave energy is absorbed. If the oscillating electric field is perpendicular to the long molecules, the wave passes through easily.

Let us call the direction in which a wave travels the x-axis and let us use the symbol y to designate the size of the wave disturbance at a position x at a given time t. In a water wave, y would represent the vertical displacement ($+$up, $-$down) of the particle at x from its equilibrium position (the position it would have if the water were perfectly flat). Obviously y varies with both position x and time t. The relation of y to x and t may be shown on a series of graphs such as Figure A3-8(1) and A3-8(2). In these graphs and in our subsequent discussion we are assuming that the waves are sinusoidal in shape. Although other shapes are commonly found in nature, the sinusoidal waves are the simplest and more important. It is easy to show that any periodic wave can be generated by the superposition of sinusoidal waves. Thus we will consider sinusoidal waves as our simple and fundamental model wave.

Figure A3-8(1) is a graph of y versus position x at one particular instant in time. We might call this time $t = 0$. The wave is sinusoidal in shape and repeats itself after a particular length called the wavelength of the wave. In Figure A3-8 the wavelength is given the symbol λ (lambda) and is the distance between two successive crests or two troughs.

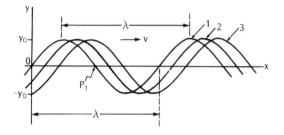

Figure A3-8
Sinusoidal wave motion

If the wave is travelling in the $+x$ direction, with the crests moving along with some speed v, then the graphs of y versus x will change with time, the crests moving to the right along the x axis. Thus curve (2) of Figure A3-8 could represent the wave at some short time after $t = 0$ and curve (3) a short time later than (2).

As the sinusoidal wave moves along, the disturbance y at any particular point (x) such as point P_1 in Figure A3-8 oscillates about its mean position in simple harmonic motion. The variation of y with time for the point P_1 of Figure A3-8 is given in Figure A3-9. The curves for other particles would be similar but shifted along the time axis. Each particle oscillates through one complete cycle in a period T as shown in Figure A3-9 and this is called the period of the wave. The reciprocal of the period, i.e. the frequency of oscillation of any particle is called the frequency of the wave. During one period T, the wave must also move forward one wavelength λ, so the frequency of the wave is also the number of wave crests (or troughs) which pass a given point per second. It follows from this that the wave

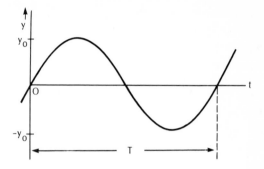

Figure A3-9
Simple harmonic motion of a particle

velocity v, wavelength λ, period T and frequency ν are related by:

$$v = \frac{\lambda}{T} = \lambda\nu \qquad \text{(A3-9)}$$

The velocity of a wave depends largely on the properties of the medium through which it moves. Thus sound waves travel at a particular speed in air at a given temperature but at a different speed in water. In addition, the wave speed may vary slightly with the wave frequency. For example, light waves of different frequencies travel at different speeds in matter although in vacuum light waves of all frequencies travel at the same speed ($3.00 \times 10^8 \text{ m s}^{-1}$). The variation of wave speed with frequency is known as dispersion and causes the spreading of "white" light into its component colours (frequencies) in a glass prism or a rainbow.

The frequency and period of a wave are determined by the wave source. Once a wave is "launched," its frequency normally remains constant, the velocity and hence the wavelength (as given by Equation (A3-9)) changing as the wave passes from one medium to another.

The sinusoidal relationship between the disturbance y, the position x and the time t, illustrated graphically by the various curves of

Figure A3-8 and A3-9, can of course, be expressed analytically. If we take the condition for $t = 0$ used in Figure A3-8 ($t = 0$ when at $x = 0$, (i) the disturbance $y = 0$ and (ii) the slope of the wave is positive), then it is easy to show that the equation for a sinusoidal wave travelling in the $+x$ direction is:

$$y = y_o \sin \left(\frac{2\pi}{\lambda} x - \frac{2\pi}{T} t \right) \quad \text{(A3-10)}$$

whereas for a wave travelling in the $-x$ direction the equation is:

$$y = y_o \sin \left(\frac{2\pi}{\lambda} x + \frac{2\pi}{T} t \right) \quad \text{(A3-11)}$$

The quantity $2\pi/\lambda$ in Equation (A3-11) is called the "wave vector" and is usually represented by the symbol "k" i.e. $k = 2\pi/\lambda$. The dimensions of k are L^{-1} and its function is analogous to the angular frequency ($\omega = 2\pi/T$) met earlier. When multiplied by the position x, the wave number generates a dimensionless quantity ($2\pi x/\lambda$) which may be considered an angle in radians. In terms of k and ω Equation (A3-9) can be expressed in the form $v = \omega/k$ or $\omega = kv$.

Equations (A3-10) and (A3-11) may be written in numerous other forms, for example:

$$y = y_o \sin \left(\frac{2\pi}{\lambda} x \pm 2\pi v t \right)$$

$$y = y_o \sin (kx \pm \omega t) \quad \text{(A3-12)}$$

$$y = y_o \sin (x \pm vt) k$$

Furthermore the travelling wave equation is often written with the x and t terms in reverse order from Equation (A3-12). For example:

$$y = y_o \sin \left(\frac{2\pi}{T} t - \frac{2\pi}{\lambda} x \right) \quad \text{(A3-13)}$$

represents a wave travelling in the $+x$ direction just as does Equation (A3-10). Equation (A3-13) simply represents a different choice of $t = 0$ than was used in Equation (A3-10).

The convention of Equation (A3-13) is followed in Chapter 2 and will for convenience be used in Section A3-4. If the same starting times are used in both Equations (A3-10) and (A3-13) then both waves would advance to the right but be 180° out of phase.

The equation of y vs. x for any given time is obtained by substitution of the given time into Equation (A3-10). For example the relation between y and x at $t = 0$ (the equation of curve (1) in Figure A3-8) is:

$$y = y_o \sin kx$$

EXAMPLE

A wave moves along a string in the $+x$ direction with a speed of 8 m s^{-1}, a frequency of 4 hertz and an amplitude of 0.05 m.

(a) What is (i) the wavelength (ii) the wave number (iii) the period and (iv) the angular frequency?

(i) Since $v = \lambda \nu$, $\lambda = \dfrac{v}{\nu} = \dfrac{8 \text{ m s}^{-1}}{4 \text{ s}^{-1}} = 2 \text{ m}$

(ii) $k = \dfrac{2\pi}{\lambda} = \dfrac{2\pi}{2 \text{ m}} = \pi \text{ m}^{-1}$

(iii) $T = \dfrac{1}{\nu} = \tfrac{1}{4} \text{ s}$

(iv) $\omega = 2\pi\nu = 8\pi \text{ s}^{-1}$

(b) What is the equation for this wave?

Since the general equation for a wave travelling in the $+x$ direction is:

$$y = y_o \sin (kx - \omega t)$$

the equation for this particular wave is:

$$y = 0.05 \sin (\pi x - 8\pi t)$$

where y and x are in metres and t is in seconds.

(c) Sketch graphs of y vs. x for this wave for the times $t = 0$ and $t = \tfrac{1}{16}$ s. State the equation relating y and x for each of these times.

(i) The equation obtained by substituting $t = 0$ into the general equation is:

$$y = 0.05 \sin(\pi x).$$

The graph of this relation is shown below.

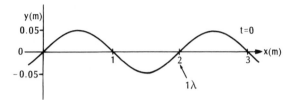

(ii) The equation obtained by substituting $t = \frac{1}{16}$ into the general equation is: $y = 0.05 \sin(\pi x - \pi/2)$. One could plot this equation by substituting several values of x and calculating the corresponding y values. However it can be sketched directly if one realizes that after $\frac{1}{16}$ s (which is $\frac{1}{4}$ of the wave period of $\frac{1}{4}$ s) the wave moves $\frac{1}{4}$ of 1 wavelength (i.e. $\frac{1}{4} \times 2 = 1/2$ m) to the right relative to its position at $t = 0$. Therefore the graph of y vs. x at this instant is:

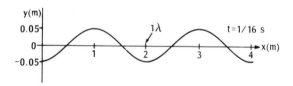

(d) What is the value of y for the point at $x = \frac{3}{4}$ m at the time $t = \frac{1}{8}$ s?

Since $y = 0.05 \sin(\pi x - 8\pi t)$

$\qquad = 0.05 \sin(\pi(\frac{3}{4}) - 8\pi(\frac{1}{8}))$

$\qquad = 0.05 \sin\left(\dfrac{3\pi}{4} - \pi\right)$

$\qquad = 0.05 \sin(-\frac{1}{4}\pi)$

$\qquad = -0.05 \sin 45°$

$\qquad = -0.035$ m

EXAMPLE

The equation of a travelling wave is $y = 4 \sin(3x + 2t)$ where y and x are in metres and t in seconds. What is the (i) amplitude, (ii) wavelength, (iii) period and frequency and (iv) velocity (speed and direction) of this wave.

By comparison with the general wave equation $y = y_o \sin(kx - \omega t)$ it follows that:

(i) $y_o = 4$ m

(ii) $k = 3$ m$^{-1} = \dfrac{2\pi}{\lambda}, \therefore \lambda = \dfrac{2\pi}{3}$ m $= 2.09$ m

(iii) $\omega = 2$ s$^{-1} = \dfrac{2\pi}{T}, \therefore T = \dfrac{2\pi}{2}$ s $= 3.14$ s

and $\nu = \dfrac{1}{T} = \dfrac{1}{\pi}$ s$^{-1} = 0.318$ s^{-1}

(iv) $v = \lambda\nu = \left(\dfrac{2\pi}{3} \text{ m}\right)\left(\dfrac{1}{\pi} \text{s}^{-1}\right) = \frac{2}{3}$ m s^{-1}

The $+$ sign in the equation indicates that the wave is travelling in the negative x-direction.

A3-4   Standing Waves

The last kind of wave we need to consider is called a stationary or standing wave. These waves are fairly common although perhaps not always obvious. They are found, for example, on guitar strings, in organ pipes and sometimes in buildings, although in the latter case the amplitude of the vibration is very small. There are many places where standing waves are important in biological systems.

The origin of the standing wave stems from the fact that when a travelling wave is reflected from a surface it suffers a phase change of 180 degrees. Thus if

$$y_1 = y_o \sin(\omega t - kx)$$

is a wave travelling to the right, then on reflection it becomes

$$y_2 = -y_o \sin(\omega t + kx)$$

a wave travelling to the left. These two waves add together to produce the standing wave. Thus

$$y = y_1 + y_2 = y_o \left(\sin(\omega t - kx) - \sin(\omega t + kx) \right). \tag{A3-14}$$

Using the standard trigonometric identity, Equations (A1-17),

$$-2 \cos \theta \sin \phi = \sin(\theta - \phi) - \sin(\theta + \phi) \tag{A3-15}$$

and Equation (A3-14) becomes

$$y = -[2 y_o \cos \omega t] \sin kx \tag{A3-16}$$

Such a wave is diagrammed in Figure A3-10 for a case where the period is 12 seconds. The shape of the wave is shown second by second over the whole period. The solid curve in Figure A3-10 represents the function at a particular instant in time. The broken lines represent subsequent positions at time intervals $\Delta t = 1,2,3 \ldots$ seconds later.

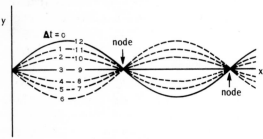

Figure A3-10

A standing wave

Typical examples of standing waves on a string are shown in Figure A3-11.

As in a guitar, the string is fixed at both ends. In this case there are two basic characteristics:

(i) Nodes (which are points where no motion is present) must be present at the reflectors or places where the string is attached.

(ii) In order to obtain a standing wave, the length of the string ℓ must equal an integral number of half wavelengths.

Thus

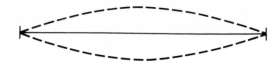

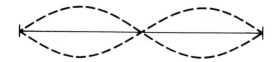

Figure A3-11

Several standing waves on a string

$$\ell = n\lambda/2$$

where

$$n = 1, 2, 3, 4 \ldots \tag{A3-17}$$

In organ pipes which are open at one end, a standing sound wave is obtained if the length of the pipe, ℓ, is equal to an odd number of quarter wavelengths. A node must be present at the closed end. Thus for this situation

$$\ell = n\lambda/4$$

where

$$n = 1, 3, 5 \ldots$$

If standing waves are produced on a ring, such as in a bell, then the circumference of the

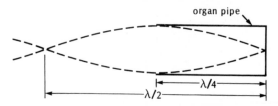

Figure A3-12

Standing wave in an organ pipe

ring must equal a whole number of wavelengths. Thus for this case

$$2\pi r = n\lambda$$

where

$$n = 1, 2, 3 \ldots \qquad (A3\text{-}18)$$

or

$$r = \frac{n\lambda}{2\pi}$$

where r is the radius of the ring.

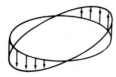

Figure A3-13
Standing wave on a ring

EXAMPLE

List three possible wavelengths for standing waves on a string of length 2 metres fixed at both ends.

$$\ell = n\lambda/2 \quad \text{or} \quad \lambda = \frac{2\ell}{n}$$

(a) $\quad \lambda = \left(\dfrac{2}{1}\right)(2)$

$\qquad = 4$ m for $n = 1$

(b) $\quad \lambda = \left(\dfrac{2}{2}\right)(2)$

$\qquad = 2$ m for $n = 2$

(c) $\quad \lambda = \left(\dfrac{2}{3}\right)(2)$

$\qquad = 4/3$ m for $n = 3$

A3-5 Beats

One last concept in waves and wave motion which we should not overlook is the process of beating by waves of two or more frequencies.

This effect can be readily experienced in sound waves and is of value in tuning pianos and other instruments. The series of clicks produced by the cricket is another sound which is thought to arise from the beating of two higher frequency sounds.

To see how this beating process occurs imagine that two waves of slightly different frequency are being received by your ear. These two waves are shown in Figure A3-14a,b.

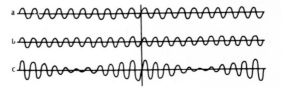

Figure A3-14
The beat pattern produced by two waves of a slightly different frequency

Since your ear responds to the total sound amplitude arriving at the eardrum (see Chapter 2) you are really experiencing the sum of these two waves, which is shown in Figure A3-14c. Because of the slight difference in frequency, the waves reinforce each other and interfere with each other at regular intervals. Thus a new frequency is produced. The value of this new beat frequency, ν_B, is the difference in frequency of the two waves arriving at the ear. Thus

$$\nu_B = \nu_2 - \nu_1$$

where ν_2 and ν_1 are the frequencies of the incoming waves. Demonstrations of beats are most impressive and the student is urged to cajole his instructor into setting up an experimental demonstration of beats using tuning forks, organ pipes or audio oscillators.

Problems

1. Consider the travelling wave: $y = 2$ sin $(\pi x + 3t)$, where x and y are measured in

metres, and t is measured in seconds.

What is (a) the amplitude?
   (b) the wavelength?
   (c) the period?
   (d) the velocity? (magnitude and direction).

2. Show that the travelling wave $y = y_o \sin (kx - \omega t)$ can be written in the alternative forms.

$$y = y_o \sin k(x - vt) \qquad y = y_o \sin 2\pi \left(\frac{x}{\lambda} - vt\right)$$

$$y = y_o \sin \omega \left(\frac{x}{v} - t\right) \qquad y = y_o \sin 2\pi \left(\frac{x}{\lambda} - \frac{t}{T}\right)$$

where y_o—amplitude
   ω—angular frequency
   T—period
   ν—frequency
   λ—wavelength
   v—velocity

3. A wave moves along a string in the positive x direction at a speed of 8 cm s^{-1} with a period of 0.25 s and an amplitude of 10 cm.
   (a) Determine the frequency, angular frequency and wavelength of the wave.
   (b) What is the equation of the wave?
   (c) Plot a graph of the wave equation as a function of x for $t = 1/6$ s.

4. Sketch the profile of the standing wave given by the equation $y = [3 \cos 2t] \sin x$ at times $t = 0$ and $t = T/4$ where T is the period of oscillation. (t in s, x in m).

5. The function $y = 7 \cos (4t) \sin (\pi x)$ describes a standing wave. If t is measured in s and x in m calculate;
   (a) the amplitude of the wave
   (b) the period of the wave
   (c) the wavelength

6. Two wave trains of the same frequency, speed and amplitude are travelling in opposite directions along a string

$$y_1 = 2 \sin (3t - x)$$
$$y_2 = -2 \sin (3t + x)$$

Find the equation of the resultant standing wave and plot the three wave forms at intervals of one-quarter of a period.

7. Two tuning forks are set side by side and struck. One fork has a frequency of 300 Hz. The other's frequency is unknown. Five beats per second are heard. The same fork of unknown frequency is now struck along with a fork of frequency 290 Hz. Five beats per second are heard again. What is the unknown frequency?

Appendix 4
Units and Symbols

The following tables contain lists of fundamental and derived units, the physical quantities and physical constants used in this book. Because there is only a finite number of letters in the Greek and English alphabets and to be in accord with common usage it is sometimes necessary to use the same symbol more than once. Certain dimensionless quantities such as the radian and the cycle are not, strictly speaking, units. Since they are frequently encountered such terms have been included but they have been placed between parentheses to remind the student that they are dimensionless. In some cases where common usage is not fully consistent with the latest S.I. recommendations an alternate set of units is given.

A4-1 Units

I. FUNDAMENTAL UNITS

Symbol	Name of Unit
m	metre
kg	kilogram
s	second
A	ampere
K	kelvin
mol	mole

II. DERIVED UNITS

Symbol	Name of Unit	Definition of Unit
C	coulomb	A s
Hz	hertz	$(cycles)\,s^{-1}$
J	joule	$kg\,m^2\,s^{-2} = N\,m$
N	newton	$kg\,m\,s^{-2}$
Pa	pascal	$kg\,m^{-1}\,s^{-2} = N\,m^{-2}$
S	siemen	$s^3\,A^2\,kg^{-1}\,m^{-2} = \Omega^{-1}$
V	volt	$kg\,m^2\,s^{-3}\,A^{-1} = J\,A^{-1}\,s^{-1}$
W	watt	$kg\,m^2\,s^{-3} = J\,s^{-1}$
Ω	ohm	$kg\,m^2\,s^{-3}\,A^{-2} = V\,A^{-1}$

III. DIMENSIONLESS QUANTITIES
radian
disintegrations
cycles
decibel (db)

A4-2 Quantities

Symbol	Name of Quantity	Units of Quantity
a	acceleration	$m\,s^{-2}$
a	average number of events (in Poisson distribution)	
A	area	m^2
A	atomic mass number	
$\mathscr{A}$	absorbance	
c	concentration	$mol\,l^{-1}$
C	concentration	$mol\,l^{-1}$
C	specific heat capacity	$J\,kg^{-1}\,K^{-1}$
d	distance between lattice planes	m
D	diffusion constant	$m^2\,s^{-1}$
E	electric field intensity	$N\,C^{-1}$
E_K	kinetic energy	J
E_n	energy of a π electron in the nth molecular orbital	J
E_p	kinetic energy of an air molecule	J
E_P	mechanical potential energy	J
E_v	energy of the vth vibrational level	J
E_ρ	total energy in a sound wave	J
f	focal length	m

227

Symbol	Name of Quantity	Units of Quantity	Symbol	Name of Quantity	Units of Quantity
F	force	N	P_x	probability density	m^{-1}
g	acceleration due to gravity	$m\,s^{-2}$	$P_{\Delta x}$	probability	
g	electric conductance	S or Ω^{-1}	q	electric charge	C
g_m	membrane conductivity	$S\,m^{-2}$ or $\Omega^{-1}\,m^{-2}$	q	image distance	m
			Q	volume flow rate	$m^3\,s^{-1}$
G	shear modulus	$N\,m^{-2}$	Q_α	energy released in alpha decay	J or MeV
H	heat energy	J			
I	current	A	Q_β	energy released in beta decay	J or MeV
I	intensity	$W\,m^{-2}$			
I	moment of inertia	$kg\,m^{-2}$	r	radius	m.
I	number of inactivating events per unit volume	(events) m^{-3}	R	radius of a blood vessel or centrifuge rotor	m
I_d	current density	$A\,m^{-2}$ $A\,cm^{-2}$	R	electrical resistance	Ω
			R	Reynolds number	
I_l	longitudinal current density in an axon	$A\,m^{-2}$ $A\,cm^{-2}$	R_l	longitudinal resistivity (in an axon)	Ω m or Ω cm
I_m	current density across a membrane	$A\,m^{-2}$ $A\,cm^{-2}$	R_m	membrane resistivity	$\Omega\,m^{-2}$ or $\Omega\,cm^{-2}$
J	flux	$mol\,m^{-2}\,s^{-1}$	s	displacement	m
k	thermal conductivity	$W\,m^{-1}\,K^{-1}$	s	sedimentation coefficient	s
k	force constant	$N\,m^{-1}$	S	sum of all spin quantum numbers	
k	wave vector	m^{-1}			
k	rate constant	s^{-1} (or min^{-1} or hr^{-1})	t	time	s
			T	period	s
			T	temperature	K
K	dielectric constant		T	transmittance	
L	intensity level	(db)	T_b	biological half-life of an isotope	s
L_f	latent heat of fusion	$J\,kg^{-1}$			
L_v	latent heat of vaporization	$J\,kg^{-1}$	T_p	physical half-life of an isotope	s
ℓ	solution thickness	m or cm			
ℓ	length of a conjugated chain	m or nm	v	velocity	$m\,s^{-1}$
m	mass	kg	v	vibration quantum number	
m	magnification of a simple lens		v_i	instantaneous velocity of an air particle	$m\,s^{-1}$
M	magnification		V	volume	m^3
M	molecular weight	$g\,mol^{-1}$	V	electrical potential difference	V
M	moment of a force	N m			
M	multiplicity		V^*	spike potential	V
n	index of refraction		V_o^*	maximum value of spike potential	V
n	quantum number for electronic energy levels				
			W	work	J
N	normal force	N	x_o	amplitude	m
p	momentum	$kg\,m\,s^{-1}$	y_o	amplitude	m
p	object distance	m	Y	(Young's) modulus of elasticity	$N\,m^{-2}$
P	pressure	Pa or $N\,m^{-2}$	Z	atomic number	
P	heat flow	W	α	angular acceleration	(radians) s^{-2}
P	power	W	γ	surface tension	$N\,m^{-1}$ $J\,m^{-2}$
P	power of a lens	m^{-1}			
$P(n)$	probability that n events will occur		ε	extinction coefficient	$l\,mol^{-1}\,cm^{-1}$
			ε	strain	

ε_s	shear strain	
η	viscosity	$N\,s\,m^{-2}$ or Pa s
θ	angular displacement	(radians)
θ_c	critical angle	(radians)
λ	wavelength	m or nm
λ_b	biological decay constant of an isotope	s^{-1}
λ_m	wavelength in a medium	m or nm
λ_p	physical decay constant of an isotope	s^{-1}
μ	linear attenuation coefficient	m^{-1}
μ	reduced mass	kg
μ_k	coefficient of kinetic friction	
μ_m	mass attenuation coefficient	$m^2\,kg^{-1}$
μ_s	coefficient of static friction	
ν	frequency	Hz or s^{-1}
ρ	density	$kg\,m^{-3}$
σ	conductivity of a pore	$S\,m^{-2}$ or $\Omega^{-1}\,m^{-2}$
σ	stress	$N\,m^{-2}$
σ_s	shear stress	$N\,m^{-2}$
τ	diffusive step time	s
ϕ	quantum yield	
ψ	wave function	$m^{-1/2}$
ψ_{cn}	cosine wave function for a ring system	
ψ_{sn}	sine wave function for a ring system	
ω	angular frequency	(radians) s^{-1}

A4-4 Some Useful Conversion Factors

Symbol	Name	Equivalent
eV	1 electron volt	$=1.6\times10^{-19}$ J
l	1 litre	$=10^{-3}$ m^3
	1 calorie	$=4.186$ J
	1 inch	$=2.54\times10^{-2}$ m
	$\log_{10} X$	$=0.4343\,ln\,X$
	$ln\,X$	$=2.303\,\log_{10} X$
°C	1 Celsius degree	$=(273.15+°C)$ K

A4-5 Prefixes

Multiple	Prefix	Symbol
10^6	mega	M –
10^3	kilo	k –
10^2	hecto	h –
10	deka	da –
10^{-1}	deci	d –
10^{-2}	centi	c –
10^{-3}	milli	m –
10^{-6}	micro	μ –
10^{-9}	nano	n –
10^{-12}	pico	p –

A4-3 Useful Physical Constants

Name	Symbol	Value
acceleration due to gravity	g	$9.8\,m\,s^{-2}$
atomic mass unit	m_u	1.66×10^{-27} kg
Avogadro's number	N_A	$6.02\times10^{23}\,mol^{-1}$
Boltzmann's constant	k	$1.38\times10^{-23}\,J\,kg^{-1}$
coulomb force constant	k	$9\times10^9\,N\,m^2\,C^{-2}$
curie	Ci	3.7×10^{10} (disintegrations) s^{-1}
electronic charge	e	1.6×10^{-19} C
electronic mass	m_e	9.1×10^{-31} kg
exponential e	e	2.71828
faraday	$\mathscr{F}$	$9.65\times10^4\,C\,mol^{-1}$
Gas constant	R	$8.3\,J\,mol^{-1}\,K^{-1}$
Planck's constant	h	6.63×10^{-34} J s
rad	rad	$0.01\,J\,kg^{-1}$
roëntgen	R	1.6×10^{15} (ion pairs) kg^{-1}

Bibliography

In this text we have covered many topics in the realm of biophysics. Of necessity the treatment is, in many instances brief. The following list of texts and articles will provide the reader with a more detailed treatment of various aspects of the material in each chapter. The reader is urged to pursue his interests by investigating these references.

Chapter 2

von Békésy, George, "The Ear," *Scientific American*, 197 (1957), 66.

Griffin, Donald R., "The Navigation of Bats," *Scientific American*, 183 (1950), 52.

Griffin, Donald R., *Listening in the Dark*, New York: Dover Publications, Inc., 1974.

Littler, T. S., *The Physics of the Ear*. Oxford: Pergamon Press, 1965.

Novick, Alvin and Bruce Dale, "Bats Aren't All Bad," *National Geographic*, 143 (1973), 615.

Pollak, George and O. W. Henson, Jr., "Cochlear Microphonic Audiograms in the 'Pure Tone' Bat Chilonycteris Parnellii Parnellii," *Science*, 176 (1972), 66.

Roeder, Kenneth D., "Moths and Ultrasound," *Scientific American*, 212 (1965), 94.

Rosenzweig, Mark R., "Auditory Localization," *Scientific American*, 205 (1961), 132.

Weidner, Richard T. and Robert L. Sells, *Elementary Classical Physics*, Vol. 1 (2nd ed.). Boston: Allyn and Bacon, Inc., 1973.

Chapter 3

Pettigrew, John D., "The Neurophysiology of Binocular Vision," *Scientific American*, 227 (1972), 84.

Sears, F. W., *Optics*. Reading, Massachusetts: Addison-Wesley, Inc., 1949.

Southwell, James P. C., ed., *Helmholtz's Treatise on Physiological Optics*. New York: Dover Publications, Inc., 1962.

Chapter 4

Blum, Harold F., *Photodynamic Action and Diseases Caused by Light*. New York: Hafner Publishing Co., 1964.

Culvert, J. G. and J. N. Pitts, Jr., *Photochemistry*. New York: Wiley, 1966.

Deering, R. A., "Ultraviolet Radiation and Nucleic Acid," *Scientific American*, 207 (1962), 135.

Jaffe, H. H. and M. Orchin, *Theory and Applications of Ultraviolet Spectroscopy*. New York: Wiley, 1962.

Swanson, Carl P. (ed.) *An Introduction to Photobiology*. Englewood Cliffs, N. J.: Prentice-Hall, Inc., 1969.

Chapter 5

Cornsweet, Tom N., *Visual Perception*. New York: Academic Press, 1970.

Gregory, R. L., *Eye and Brain*. New York: World University Library, McGraw-Hill, 1970.

Chapter 6

Andrews, Howard L., *Radiation Biophysics* (2nd ed.). Englewood Cliffs, N.J.: Prentice-Hall, Inc., 1974.

Casarett, Alison P., *Radiation Biology*. Englewood Cliffs, N. J.: Prentice-Hall, Inc., 1968.

Chapter 7

Alexander, R. McNeill, *Animal Mechanics.* Seattle: University of Washington Press, 1968.

Benedek, George B. and Felix M. H. Villars, *Physics with Illustrative Examples from Medicine and Biology*, Vol. 1 *Mechanics.* Reading, Massachusetts: Addison-Wesley, Inc., 1973.

Dyson, Geoffrey, *The Mechanics of Athletics.* London: University of London Press Ltd., 1970.

Hildebrand, Milton, "How Animals Run," *Scientific American*, 202 (1960), 148.

Napier, John, "The Antiquity of Human Walking," *Scientific American*, 216 (1967), 56.

Thompson, D'Arcy, *On Growth and Form* (Abridged edition edited by J. T. Bonner). London: Cambridge University Press, 1971.

Sutton, Richard M., "Two Notes on the Physics of Walking." *American Journal of Physics*, 23 (1955), 490.

Weidner, Richard T. and Robert L. Sells, *Elementary Classical Physics*, Vol. 1 (2nd ed.), Boston: Allyn and Bacon, Inc., 1973.

Yamada, H., *Strength of Biological Materials.* (ed. F. G. Evans), Baltimore: Williams and Wilkins Co., 1970.

Chapter 8

Dickerson, R. E. and I. Geis, *The Structure and Action of Proteins.* New York: Harper and Row, Publishers, 1969.

Perutz, M. F., "The Hemoglobin Molecule," *Scientific American*, 196 (1964).

Phillips, D. C., "The Three-Dimensional Structure of an Enzyme Molecule," *Scientific American*, 215 (1966) 78.

Setlow, Richard B. and Ernest C. Pollard, *Molecular Biophysics.* Reading, Massachusetts: Addison-Wesley, Inc., 1962.

Snell, F. M., S. Shulman, R. P. Spencer and C. Moss, *Biophysical Principles of Structure and Function.* Reading: Addison-Wesley Publishing Co. Inc., 1965.

Watson, James D., *Molecular Biology of the Gene.* New York: W. A. Benjamin, Inc., 1965.

Wilson, H. R., *Diffraction of X-rays by Proteins, Nucleic Acids and Viruses.* London: Edward Arnold Publishers Ltd. 1966.

Chapter 9

Alexander, R. McNeill, *Animal Mechanics.* Seattle: University of Washington Press, 1968.

Evans, L., "Blowing Up Baby," *The Sciences*, 12 (1972), 26.

Nobel, Park S., *Introduction to Biophysical Plant Physiology.* San Francisco: W. H. Freeman and Co., 1974.

Warren, J. V., "The Physiology of the Giraffe," *Scientific American*, 231 (1974), 96.

Chapter 10

Crawford, Franzo H., *Heat, Thermodynamics and Statistical Physics.* New York: Harcourt, Brace & World, Inc., 1963.

Van Holde, K. E., *Physical Biochemistry.* Englewood Cliffs, N. J.: Prentice-Hall, 1971.

Reif, F., *Fundamentals of Statistical and Thermal Physics.* New York: McGraw-Hill, Inc., 1965.

Setlow, Richard B. and Ernest C. Pollard, *Molecular Biophysics.* Reading, Massachusetts: Addison-Wesley, Inc., 1962.

Tanford, C., *Physical Chemistry of Macromolecules.* New York: Wiley, 1961.

Chapter 11

McDonald, D. A., *Blood Flow in Arteries.* London: Edward Arnold, 1974.

Weidner, Richard T. and Robert L. Sells, *Elementary Classical Physics*, Vol. 1 (2nd ed.). Boston: Allyn and Bacon, Inc., 1973.

Scott Blair, G. W., *An Introduction to Biorheology.* New York: Elsevier Scientific Publishing Company, 1974

Chapter 12

Burton, A. C. and O. G. Edholm, *Man in a Cold Environment.* London: Edward Arnold, 1955.

Carey, F. G., "Fishes with Warm Bodies," *Scientific American,* 228 (1973), 36.

Monteith, J. L., *Principles of Environmental Physics.* New York: American Elsevier, 1973.

Chapter 13

Cole, K. S. *Membranes, Ions and Impulses.* Los Angeles: University of California Press, 1972.

Cooke, I. and M. Lipkin, Jr. *Cellular Neurophysiology.* New York: Holt, Rinehart and Winston, Inc., 1972.

Katz, B., *Nerve, Muscle and Synapse.* New York: McGraw-Hill Book Company, 1966.

Wilson, J. A., *Principles of Animal Physiology.* New York: The Macmillan Company, 1972.

Answers to Problems

Note: The answers to many problems have been calculated using $g = 10\text{ m s}^{-2}$.

CHAPTER 2

1. $35\text{ m}, 0.10\text{ s}, 20\pi\text{ s}^{-1}$
2. (a) 0.12 W
   (b) $2.4 \times 10^{-4}\text{ J}$
   (c) 0.12 W through each sphere
   (d) at $r_1 = 1\text{ m }I_1 = 10^{-2}\text{ W m}^{-2}$
      at $r_2 = 5\text{ m }I_2 = 4 \times 10^{-4}\text{ W m}^{-2}$
   (e) $4 \times 10^{-9}\text{ W}, 8 \times 10^{-12}\text{ J}$
   (f) 4.3 nm.
3. (a) $6.3 \times 10^{-10}\text{ J}$
   (b) $8.3 \times 10^{-9}\text{ m}$
4. $8.9 \times 10^2\text{ m}$
5. $5.08 \times 10^{-2}\text{ m}, 6.7 \times 10^3\text{ Hz}$
6. 3.9 m (this distance is very large; therefore some other device must be used).
7. 103 db
8. 3 db
9. (a) yes
   (b) no
10. $2.3 \times 10^5\text{ W}$
11. $100\text{ db}, 40\text{ db}$
12. 80 db
13. (a) 12.6 m
    (b) $1.2 \times 10^{-11}\text{ m}$
14. 6.3 m s^{-1}
15. 0.85 m

CHAPTER 3

1. $4.6 \times 10^{-7}\text{ m}, 4.8 \times 10^{14}\text{ Hz}, 2.2 \times 10^8\text{ m s}^{-1}$
3. $2.1 \times 10^8\text{ m s}^{-1}$
4. 16 times
5. 10^{14}
6. (a) -25 d
   (b) -8.5 d

(c) $+8\text{ d}$
(d) $-0.04\text{ m}, -0.04\text{ m}; -0.16\text{ m}, -0.16\text{ m}; 0.13\text{ m}, 0.17\text{ m}$
(e) 0.061 m to left of lens
   0.67 m to left of lens
   11.3 d
7. 0.333 m
8. $-0.500\text{ d}, -2.00\text{ m}$
9. (a) $65\text{ d}; 5\text{ d}$
   (b) 5.5 m
10. 2 mm
11. yes
12. $-0.43\text{ m (q)}, 1.1\text{ m (inverted)}$
13. 0.27 m
    0.03 m

CHAPTER 4

1. $3.06 \times 10^{-10}\text{ m}$
2. No
3. ── ⇅ ↑ ↑ ↑ ↑
   ⇅ ── ↑ ↑ ↑ ↑
4. $7.65 \times 10^{-19}\text{ joule}$
5. $4.4\text{ nm}^{-1}, 2.2\text{ nm}^{-1}, 0.02$
6. $1.19 \times 10^{-18}\text{ joule}$
7. $\psi_2 = \sqrt{\dfrac{2}{0.45}}\sin\left(\dfrac{2\pi x}{0.45}\right); 3.33\text{ nm}^{-1}$
8. 0.7 nm^{-1}
9. $1.35\text{ nm}; \quad 9$
11. 292 nm
13. $E_e > E_v > E_r$
14. a combination of all three
15. $2, 4, 6, \ldots$
16. $2.2 \times 10^{-7}\text{ m}$
17. $2 \times 10^{-19}\text{ J (per photon)}$
18. 0.80
19. 5
20. (a) 2

(b) 3, 6, 6, 4, 3, 1
(c) reasonably good
21. 25%
22. green
23. 13%
24. 1.3, 5%

CHAPTER 5

2. (i) 0.368
   (ii) 0.264
3. 12%
4. 170
5. 20
6. 0.20

CHAPTER 6

1. (a) $^4_2$He, α particle or helium nucleus
   (b) $^0_{-1}$e β particle or electron
2. 4 days
3. (a) 49 : 1
   (b) 0.12 : 1
4. 4.3 days
5. 33 hr.
6. 0.66 days
7. (a) 0.409 cm
   (b) 200 cells
8. (a) 5×10^4 eV
   (b) 0.025 nm
   (c) 0.09 cm
9. 5×10^4 erg
10. 5 times
11. 0.10 cm^2 g^{-1}
12. 3.6 cm
13. 4.5×10^{-15} cm^3
14. 0.027
15. 6.1×10^9 events cm^{-3}
16. 0.12 or 12% hit
17. (a) 2 to 2.5
    (b) 1.8×10^{-14} cm^3

CHAPTER 7

1. 8.45 N, -18.1 N

2. (a) $(-6.0, 2.0)$ m
   (b) 6.3 m 18° N of W
3. 231 N, 116 N
4. 21 m s^{-1} north-west
5. (a) 38.7° W of S
   (b) 64 hr.
6. 11 m s^{-1}
7. 513 N m
8. 50 N
9. 6.43×10^3 N
10. $T = \dfrac{\sin \theta}{\sin \Delta}(2W + 5W_1)$
11. 1.2×10^3 N
12. $T_1 = 1125$ N, $T_2 = 625$ N
13. $h = \dfrac{v^2 \sin^2 \theta}{2g}$
14. $F_m = 1330$ N, 1230 N
15. $F_m = 1540$ N, 1450 N at 32° from vertical.
16. 7.75 m s^{-1}
17. 15 N
18. (a) 1.83 s
    (b) 15.3 m
19. 2 m s^{-2}
20. 1.6 m
21. 0.05
22. 4.9 m s^{-1}
23. (a) $W_F = 98.3$ J, $W_{F_f} = -84$ J
        $W_{mg} = 0, W_R = 0$
    (b) 14.3 J
    (c) 2.39 m s^{-1}
24. 1.01
25. 10 times larger in arthritic joint.
26. 26.6°
27. 2.4×10^3 N
28. 1.67×10^{-2} N
29. 47 m s^{-1}
30. $F_c = 2.37\, F_g$
31. Conservation of angular momentum. Since the moment of inertia is less in the tucked position, therefore the angular velocity is greater.
32. (a) 4.0 radian s^{-1}
    (b) 6.0 radian s^{-1}

33. 17 m s^{-1}
34. 7×10^{-3}
35. $7.3 \times 10^5 \text{ N m}^{-2}$
36. $6.3 \times 10^{-3} \text{ m}$
37. $6.2 \times 10^2 \text{ N m}$ (assuming modulus $= 10^{10}$ N m^{-2})
38. 10^{10} N m^{-2}
39. $d \geq 2.1 \times 10^{-3} \text{ m}$ for a 70 kg person
40. $W \propto L^3$
42. 400 gm.
43. 31.6 times that of the smaller ant.
44. 0.0185 kg; 0.7% (man), 3.7% (dog)
45. 3.4:1

CHAPTER 8

1. (a) $-5.76 \times 10^{-22} \text{ J}$
   (b) $-4.61 \times 10^{-20} \text{ J}$
2. increases by 44%. (i.e. it might change from -100 to -56).
3. (d)
4. (d)
5. (a) $0.40°, 0.80°, 0.12°$
   (b) $0.80°$
6. (a) 0.20, 0.13, 0.10, 0.80 nm
   (b) $60°$

CHAPTER 9

1. 0.37 times atmospheric pressure
2. 20 km
3. 0.37 times atmospheric pressure
4. $3.26 \times 10^{-29} \text{m}^3$
5. $2.17 \times 10^5 \text{ Pa}$
6. $1.8 \times 10^4 \text{ Pa}$ (for a 1.8 m tall person)
7. $1.6 \times 10^4 \text{ Pa}$ (a gauge pressure)
8. 1.2 m
9. gauge pressure $= 2 \times 10^3 \text{ Pa}$
   absolute pressure $= 1.02 \times 10^5 \text{ Pa.}$
10. $1.8 \times 10^5 \text{ Pa}$
11. greater by a factor of 2

12. $11 \times 10^{-3} \text{ N m}^{-1}$
13. $7 \times 10^2 \text{ Pa}$
14. (a) zero
    (b) -32 Pa
15. (c)
16. $4 \times 10^2 \text{ N}$ (for castor oil).

CHAPTER 10

1. (a) 0
   (b) $3.89 \times 10^{-4} \text{ m}^2$
   (c) $1.97 \times 10^{-2} \text{ m}$
   (d) 0
2. 3
3. $10^{-2}:1$
4. 83 s
5. $9.5 \times 10^{-6} \text{ m}$
6. $1.7 \times 10^{-2} \text{ s}$
7. 1/2 that of the smaller molecule
8. it would increase by 100 times
9. $4 \times 10^{-3} \text{ N s m}^{-2}$
10. 4×10^{12} (Wow!)
11. $1.9 \times 10^{-9} \text{ m}$
12. $9.5 \times 10^{-16} \text{ m}$
13. $0.87 \text{ m s}^{-2}, 0.58 \text{ m s}^{-1}$
14. (a) 10^6
    (b) $2.8 \times 10^3 \text{ s}$
15. 10 times
16. 10^4 kg m^{-3} (assume the liquid medium is water).
17. both increase the diffusion constant and increase the molecular weight.

CHAPTER 11

1. (b)
2. $7.5 \times 10^3 \text{ Pa}$
3. $3v_1^2/2 g$
4. $1.96 \times 10^5 \text{ Pa}$
5. $4.4 \times 10^{-3} \text{ m}$
6. (a)
7. 3×10^8
8. $(4v_s/25) \text{ m s}^{-1}$

9.

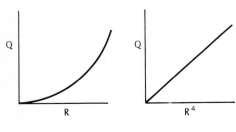

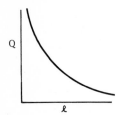
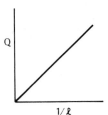

10. P diseased $= 2.78$ P healthy
11. 450; die out
12. (a)
13. (a) $\approx 8 \times 10^{-4}$ m^3 s^{-1}; ≈ 2.5 m s^{-1}
    (b) ≈ 4000; possibly turbulant
14. 0.82 N m^{-1}
15. 1.6×10^{-2} N m^{-1}
16. 0.33 cm^3; 54 cm^3.

CHAPTER 12

1. 1.5 hr
2. more
3. 0.27 C
4. 8500 K(if $\lambda = 600$ nm)
5. 0.64
6. 25.4 min
7. 0.33 kg

CHAPTER 13

1. 1.6×10^{-14} A
2. (a) 1.3×10^{-14} A
   (b) 2.6×10^{-8} A cm^{-2}
3. 5×10^{-4} A cm^{-2}, 2.5×10^{-11} A
4. (a) $0, 0, -2.2 \times 10^3$ V
   (b) -1.07×10^4 N C^{-1}, -1.07
   $\times 10^{-4}$ N, -1.09×10^3 V

5. 1.2×10^{-20} N
6. i is 14.2×10^{-3} V negative with respect to o.
7. 0.14 mol l^{-1}, 0.06 mol l^{-1}, 0.30 mol l^{-1} about 20 mV
8. -63 mV, $+58$ mV, -85 mV
10. 5.3×10^{-6} Ω^{-1}
11. (a) 62 Ω
    (b) 162 Ω
12. (a) 0.24 cm
    (b) 0.17 cm
13. 0.12 cm
14. 1.4

Answers to non-graphical problems in appendices.

APPENDIX 1: Pg. 207, PROBLEMS

1. (a) 0.76
   (b) 1.76
   (c) -0.24
   (d) -0.76
   (e) 4.31
   (f) 1.74
2. (a) -6.59
   (b) -5.47
3. (a) -0.72
   (b) -0.44
4. (a) 0.50
   (b) 0.50
   (c) 148
   (d) 2.75
5. 10
7. 1.98×10^{-2} yr^{-1}
8. 6 min.
9. 8 AM
10. 2.7 times
11. 3rd culture
12. May 6
13. 0.18 min^{-1}
16. "a" $= y -$ intercept

APPENDIX 2: Pg. 211, EXERCISE

The only incorrect one is (iii).

Pg. 212, EXERCISE

(i) $[k] = ML^3 T^{-4} A^{-2}$
(note that $[k]$ means "dimensions of k")
− units are kg m^3 s^{-2} C^{-2}
(ii) $[R] = ML^2 T^{-2} \theta^{-1}$; units are kg m^2 s^{-2} K^{-1} mol^{-1}
(iii) yes; P: atmospheres n: moles
         V: litres T: K

Pg. 213, EXERCISE

(i) $[\omega] = T^{-1}$; $[k] = L^{-1}$
(i) $[\lambda] = T^{-1}$; units of $\lambda = $ s^{-1}
(iii) 1 mol^{-1} cm^{-1}

Pg. 213, PROBLEMS

1. $ML^{-1} T^{-3}$; kg m^{-1} s^{-3} (or N s^{-1} m^{-2})
2. (b)
3. J K^{-1}

4. (a)

5. $ML^2 T^{-1}$; kg m^2 s^{-1} (or: J s)

APPENDIX 3: Pg. 225, PROBLEMS

1. (a) 2 m
   (b) 2 m
   (c) $\dfrac{2\pi}{3}$ s
   (d) $\dfrac{3}{\pi}$ m s^{-1} in the negative x-direction

3. (a) 4 s^{-1}; 8π s^{-1}; 2 cm
   (b) $y = 10 \sin(\pi x - 8\pi t)$
     (or: $y = 10 \sin(8\pi t - \pi x)$)

5. (a) 7m,
   (b) $\dfrac{\pi}{2}$ s
   (c) 2m

6. $y = -4 \cos 3t \sin x$

7. 295 Hz

Index